Pennsylvania's
Paleozoic Playground

Gastropods, from McAlisterville, Juniata Co. Pa. (Devonian Age)

Cover: **State Fossil, *Phacops Rana*** with Pelecypod, Mahantango Formation Perry Co. Pa.

Library of Congress Control Number: 2009913645
ISBN:
Softcover - 978-1-4500-1551-6
Hardcover - 978-1-4500-1552-3

This book was printed in the United States of America.

To order additional copies of this book, contact:
Xlibris Corporation
1-888-795-4274
www.Xlibris.com
Orders@Xlibris.com

Dear Readers: The book you are holding in your hands has evolved from its original state. What you have is a compilation of four individual books metamorphosed into one.

My book started as Pennsylvania's Paleozoic Playground with beginning chapters including our local Lower Cambrian and Ordovician faunas each of which later evolved into separate more detailed compilations as follows. **The Lower Cambrian Explosion of life in my back yard:** and **Swatara Gap Pa. Ordovician fossil fauna**.

Within this publication original pages on those subjects have been removed and replaced with the individual, more detailed books in their entirety. However for convenience the page number sequence of each book including indexes have been left intact for each of the newer additions. Also, I included another different book portraying the fossils of Calvert Cliffs of Maryland as a bonus for your added pleasure. My primary interest is with the local fauna. Nevertheless I have always enjoyed collecting the Calvert marine vertebrate faunas a few hours south of the border for the recreational experience, personal enjoyment of exploring, and specimen collecting. Now available to share with all of you.

.... Enjoy!

Kerry D. Matt

Pennsylvania's Paleozoic Playground

Why such a title? One might ask! Well, quite frankly, it has a nice ring to it. But isn't the playground concept a bit unorthodox regarding the science of paleontology? My reasoning is based on the following: The more kinds of recreation available to the public, the better, and shouldn't always have to be limited to sports activities. To be honest, fossil collecting is becoming more popular with the public whenever it is exposed as a hobby interest. Going into the field to look for fossils is a fun, entertaining, and healthy activity where one can enjoy the outdoors, get fresh air, plenty of exercise, and sun. Of course, one of the most important things to remember with any hobby or recreational activity involving science is education. There are professionals working hard in the disciplines of paleo-sciences who may be offended at the very idea of amateurs and novices engaging in such endeavors as extracting fossils. However there is always another argument! All of us have to start somewhere, including those reaching for the very top of the academic ladder. Where paleontology is concerned, there will never be enough academic staff, or time, to explore the ever-changing terrain of our fossil environment. In fact, the more eyes and hands involved, the more precious fossil remains can be found, saved and preserved from forces of construction, natural erosion, and disintegration due to weathering.

What is important is the educational part. Teaching paleontology, earth science, and history is essential to treating scientific hobbies with respect and discipline so such activities remain functional to all of us, as a whole. It is important to keep a log of where items are found and be sure to share valuable information with the scientific community. I always say fossils have been in the ground for millions, or hundreds of millions, of years. If a prize specimen remains in a child's drawer or desktop for a lifetime, it really doesn't matter. What is most important is, it is preserved, can be studied, if necessary, and can be put in a museum sometime later.

Fossils are mostly like photographs of life's history preserved in rock. Most times (in organic terms) there is nothing, or little left of an animal after the fossilization process has occurred, but a picture can be worth a thousand words. Often, fossils can be photographed or plaster casted (with great care) to keep valuable information flowing in palentological archives. There is nothing that is either sacred or private regarding our earth's history. Information is priceless, yet worthless, if not shared by all. What we really need are good fossil collecting parks to inspire new interests and activities for today's new generation. Good quality activities stimulating new values, and a better lifestyle. I will always remember how I felt the first time I was exposed to fossils and minerals. It certainly set a better course for my life!

Kerry Matt (The Stone Man!)

Discovery emerges from youthful curiosity . . . . defining the ever-flowing fountain of youth!

From the time I was a small child, as far back as I can remember, I was always curious about my natural surroundings. I feel this is where interest in natural science emerges for many who persevere endless enthusiasm and curiosity in cosmetology and all the byproducts thereof. Now, we are face to face with everything that makes things be. Simply put, it all starts with, "which came first, the chicken or the egg"? Or should I be saying, "physics or astronomy"? Actually, it all starts with the bonding forces of physics and chemistry. Then you have substance, matter, a universe full of galaxies, stars, planets, (astronomy). Places where chemical and physical substance can be, taking endless form, or better yet, the "stuff of the stars"! Then comes the emergence of life, a long process leading to the kinds we are familiar with, and can identify with. To pursue the natural sciences is really to find the real us, who we are, a reflection of all that is. As a philosopher, I could write a whole book on this subject, and that may come someday. To be quite honest, I believe life is more than bio-chemical compositions sustaining personality by competing, feeding, or reproducing, but rather, all things are alive, even rocks. Philosophically speaking, if it can have form or structure, it lives by the physical forces making up its character governing whatever substances are in its composition, and capable of maintaining within the environment of which it exists. Now, with all this being said, it is time to get to a point. It is about shaking down one of the natural sciences. This is paleontology, the study of the fossil record of life, its origin, history, and where we all came from, as recorded in the rock strata surrounding us

For me it all started back when an old friend from the neighborhood, just a few doors down, presented a slab of fossil ferns his father brought back from a hunting trip. I was thoroughly amused at the solidified plant embedded in this slab of rock. It seemed rather unreal at the time, and I thought, how could this be? They're in solid stone; a plant in full detail showing vein structure raised like brail yet no fleshy material there at all! Right then and there, I knew I needed to have some of this. I recall getting a small piece to take home and experiment with. I remember actually causing the material to disintegrate, because I persisted in repeatedly splitting the material, with the fascination of seeing the fossils emerge from within this rock. How could a plant end up turning to solid stone? How can a plant be in there and become like this, I thought.

In those days, other kids in the neighborhood were off playing sports and all the usual fun and games. For me, I was off in a different world, wondering through the woods, and exploring. I needed another fossil, but had no clue how to get one. I looked everywhere, but could not find a trace, not when living on top of the non-fossiliferous Conestoga Limestone Formation which, of course, I had no idea at that time. I will never forget how my friend and I, (who were always the impish types), decided to make our own fossil. We found fresh fern leaves, gathered old roofing slate, and proceeded to place the fern leaves neatly between the slate pieces, burying them near the old city dump in a woods next to our neighborhood. I remember telling him something like "we can come back and check these when we are real old like maybe when we become 30". Fascination with rocks, plants, animals, insects, reptiles, the night sky, all emerged with frequent trips to the local city museum and, from there, exploded into a new obsession. In the museum, I saw the samples for sale at the sales desk in the lobby. I spotted those fossil plants again in small plastic bags with a label. "Hey, dad? Can ya' buy me one of those! Now I was in this new game, a kid wanting to start my own "Neighborhood Museum."

Discovery emerges

Now, as for the small package of fossils we bought at the museum. Turns out, this fossil leaf, showing much detail in a hard reddish brown rock, almost looking like a big feather, was the leaf of nothing more than your common everyday *Neuropteris Scheuchzeri*, (pronounced <ner-op-ter-iss-skook-zer-i>). This fossil had a little "hidden fore-shadowy" to my future. It turned out; later in life I became good friends with a professor who had a PhD in geology. I eventually learned the story behind this particular fossil. I had no idea, at the time I bought the specimen as a kid; the fossil came from a site in Pa. It was documented as one of the finest sites for exquisite preservation in plant fossils. At that time, I barely knew my future geologist friend, who discovered the site. He was working his way through college, packaging and selling some of the smaller specimens from his magnificent find from the coal regions in northern counties, in order to help pay for his education. Ironically, through him, I was later able to acquire magnificent large-scale examples from the site, which at present no longer exists, due to mine reclamation.

It was not long, via regular weekend visits, until my dad's old music teacher, who was a volunteer at the desk of the museum, suggested I may be of age to get involved with a program, a club affair, known as the Junior Earth Science Club. I soon joined, and was welcomed to a whole new dimension to my world of interest. I was now, for the first time, able to go on field trips making hands on discoveries of fossils and minerals I could only imagine finding, after seeing them through glass in museum displays.

I was always very interested in both fossils and minerals. My first book, **Pennsylvania's Rainbow Under Ground**, documents a collectors experience with Pa. minerals. This book tells the fossil collecting part of the journey, I am still pursuing today. I have clear memories of my first fossil expedition with the museum group. It was a Lower Cambrian trilobite locality, known as Fruitville Quarry, which now stands fenced off in a suburban housing development. Franklin and Marshall College currently owns the quarry as a lab site. At one time, it was simply an open pit on the side of a knoll. You could walk a path through a small patch of woods to find an area full of exposed gray shale. There, I found my first sample of fossil trilobite. I still have the piece, a vaguely distinguishable pair of head shields. I remember at the time, how strange these appeared, having no clue just what they were called, or how this occurrence could even exist. But it sure was neat from a kid's point of view. It wasn't long before the experienced collectors were coming up with the good stuff. I had my first look at a complete trilobite found in the field. Then I was really hooked! Wow! It looks like someone took a big centipede and mashed it between two rocks! There was a big funny looking oval shaped head, crescent shaped eyes with a segmented looking skeleton-like body, and a spiny tail. In those days, for me, this was "way too cool"! I had to get one of those and learn more about this peculiar material, for sure.

There was one strange irony about these first fossil trilobites that always comes to mind, something about the structure of the eyes and the glabella (nose) within the head shield of these trilobites (*Olenellus Thompsoni*). The structure reminded me of an emblem appearing in my Boy Scout manual. I recall the days where I was not very successful in Boy Scouts. I was a bit too freelance and independent to be successful in the Scouts. That is when it was decided I needed a different outlet for my interests and goals. The introduction to the Junior Earth Since Club, all the fond memories, and help of many white-haired mentors from the museum days set the pace for my future endeavors as an amateur Paleontologist.

Paleozoic Paleontology of Pennsylvania

The study of Earth's fossil record is the science of paleontology. There were many occasions where people have mistaken me for an artifact hunter, applying to an entirely different dimension of pre-historic science, archeology, the study of pre-historic man and his position in the history of our planet. Nevertheless, there's a sort of a fuzzy gray line as to where Paleo-history and human history tend to merge. There are humans in the fossil record. In the case where Pennsylvania is concerned, the fossil history begins much earlier.

The evolution of life on this planet involves a complex series of dated time zones depicting the kinds of various species of animals existing within environments our newly evolving planet allowed. Most rock strata in Pennsylvania bearing any significant fossil record applies to a very early time period known as the Paleozoic era. This is divided into sub sequential zones, where specific paleo-biospheres allowed progress in the evolution of animals, as we know them today.

Rocks in Pennsylvania also bear sediments exposing advanced life forms, such as dinosaurs, from a time long after the much older Paleozoic era. The fossils focused on in this book mostly predate any land roaming creatures by a long shot, though creatures were beginning to emerge onto land as far back as the Devonian period within the Paleozoic era. The fossil record of Pa. concerning dinosaurs does not stand in the spotlight with high ranks as compared to other places in the world. However there is a lot of ground to cover within the Triassic/Jurassic periods, involving Mesozoic vertebrate paleontology. The facts are quite simply the rarity and scarcity of such kinds of preserved fossils to be found in Pa. competing with the fossil record over all. When found, such specimens are held with high regard because of this. I give two thumbs up to those who pursue vertebrate paleontology in the later fossil record, here in Pa.

Getting back to the Paleozoic period with Pa. demonstrates the primary and most easily accessible kinds of fossils one can expect to encounter in this geological environment. Exposure is a major factor here. If there were spectacular vertebrate sites, most would likely be hidden under our rich forests and highly cultivated lands. In places where rocks are exposed, such as the great deserts of the world, the chances are much greater in finding such sites. This is cutting edge paleontology. Here in the east, the earths exposed crust is a bit older, well weathered and hidden within our lush rich green topography. By saying this, exposure is a key ingredient in what kinds of paleontology can be pursued, and to what degree.

Paleozoic period rock is the most commonly exposed in this state. I have a theory about developing a good representation of a fauna as applied to the record. I refer to this as "concentration versus elements of shabbiness", in a humorous sense. This is why my book focuses on a limited area applying to my home state. As a fossil hunter, I know there is so much out there and so many sites to go collecting. I realize one person can only have so much time to work any given area. Probability and statistics tells me, if I visited many sites all over the country or beyond, I could only acquire limited samples from whatever beds I am digging on any particular day. Therefore, making a limited fauna representation, more or less a collection of knick-knacks!

I worked specific areas and sites over the years closer to home. This enabled me to concentrate on areas where I could statistically produce better specimens. More complete faunal assemblages within a few hours' drive, at best. I have always treated this hobby from a scientific approach and point of view. With this book, I aim to share a better picture of Paleozoic Paleontology, representing, more or less, a 100-mile stretch throughout the Susquehanna Valley of Pa. This demonstrates a good representation of what can be found in the local fossil record from earliest periods, exposing ancestral creatures similar to many living today.

This book is most useful as a specific documentary. Other books on paleontology are recommended for in-depth studies and further illumination. I can give a brief overview of how the story begins with fossils in Pa. First of all, for some one new at this, one might ask, "What is a fossil?" Where do fossils come from? To best answer this, one has to imagine going back in time. Just what did happen? And, when? Most fossils we are familiar with, originated from bodies of water. You need something of organic composition (in this case, excluding trace fossils such as footprints, burrows, etc.) buried in sediment, silt, mud, ash, etc., and the conditions have to be just right. This is another factor of probability and statistics. Remember, all fossils are rare, but keep in mind, if you find one of anything, no matter how rare the species, in life, there were most likely hundreds of millions of each of these individuals living all over the planet. What happens is, in the fossil record, we get lucky enough to stumble upon a few that were at the right place at the right time for preservation. The rest have all decayed. You need the right circumstances in the pre-existing environment, where the decay process is slowed down enough. Minerals can replace organic cell structure to completely replace the animal or plant, preserving its image and structure in rock. This is usually an environment where there is a lack of oxygen, which is a key element in rapid decay. Another issue of importance is the burial place. Whether any such animal is disturbed by scavenging protozoa, bacteria, or even physically disturbed by other animals still functioning, such as burrowing worms, mollusks, or whatever. The majority of all fossils we see have no organic matter left. Minerals replaced them, but there are a few that have been metamorphosed, where shell, tooth, or bone matter recrystalized still maintaining much of the original composition. Plant material has left a huge mark in our Paleozoic environment here in Pa. in the form of carbon remains. Carbon can be observed as residual deposits left from most fossil plants in the form of coal, and traces in ancient limestone beds where algae once existed.

How did the fossil get here? If one could turn back time and see the new earth in its stages after cooling, a shroud of water vapors eventually precipitated into a large ocean, covering most of the entire globe. One could see the original Earth was a slight bit out of round. On one side, there was a single super continent, which was the high point of the new cooling planet. Gravitational forces, over time, pulled the earth into its present round shape. The super continent breaking into individual sections drifted apart, breaking up into all the separate continents we are familiar with today as they all floated upon the liquid metal mantle of the earth. Forces pushing massive liquid rock between the plates slowly separated the landmasses. The point of my referencing this geologic past, which can be studied in textbooks as plate tectonics (continental drift), is to make one realize most fossils we find locally, actually never lived here in longitudinal and latitudinal terms. The continent of North America was in a different place on the globe having a warmer climate. It appeared much different during the Paleozoic period. There were large seas existing inland, over places where we now see mountains. Philosophically speaking, many oceans have danced across the face of what Pa. once was during the slow journey to the present day position on the globe. With all those hundreds of millions of years, many different life forms were buried in sediments. Ancient seabeds were later transformed into mountains by the great forces, of plate tectonics. There was once a time when the continent of Africa crashed into and later pulled away from North America folding rock layers into our familiar mountains and hills, during contact, then creating a great rift after pulling away. You can observe layers of sediment and original bedding planes, which were in many places, forced into a vertical position over eons of time. The study of deposited sedimentary layers is a science called stratigraphy. Great mountain ranges once standing east of Pa. filled in original ocean basins, and the great rift as well, producing sedimentary layers observable in outcrops along our roadsides and cuts. Fossils each layer could bear are absolutely predictable, because stratigraphy tells a story of the time they were deposited, and the kinds of animals living during a particular period. I hope this publication can sort out a few of these for your closer observation.

In My Back Yard!

By: Kerry Matt

Lower Cambrian Explosion of life in my backyard

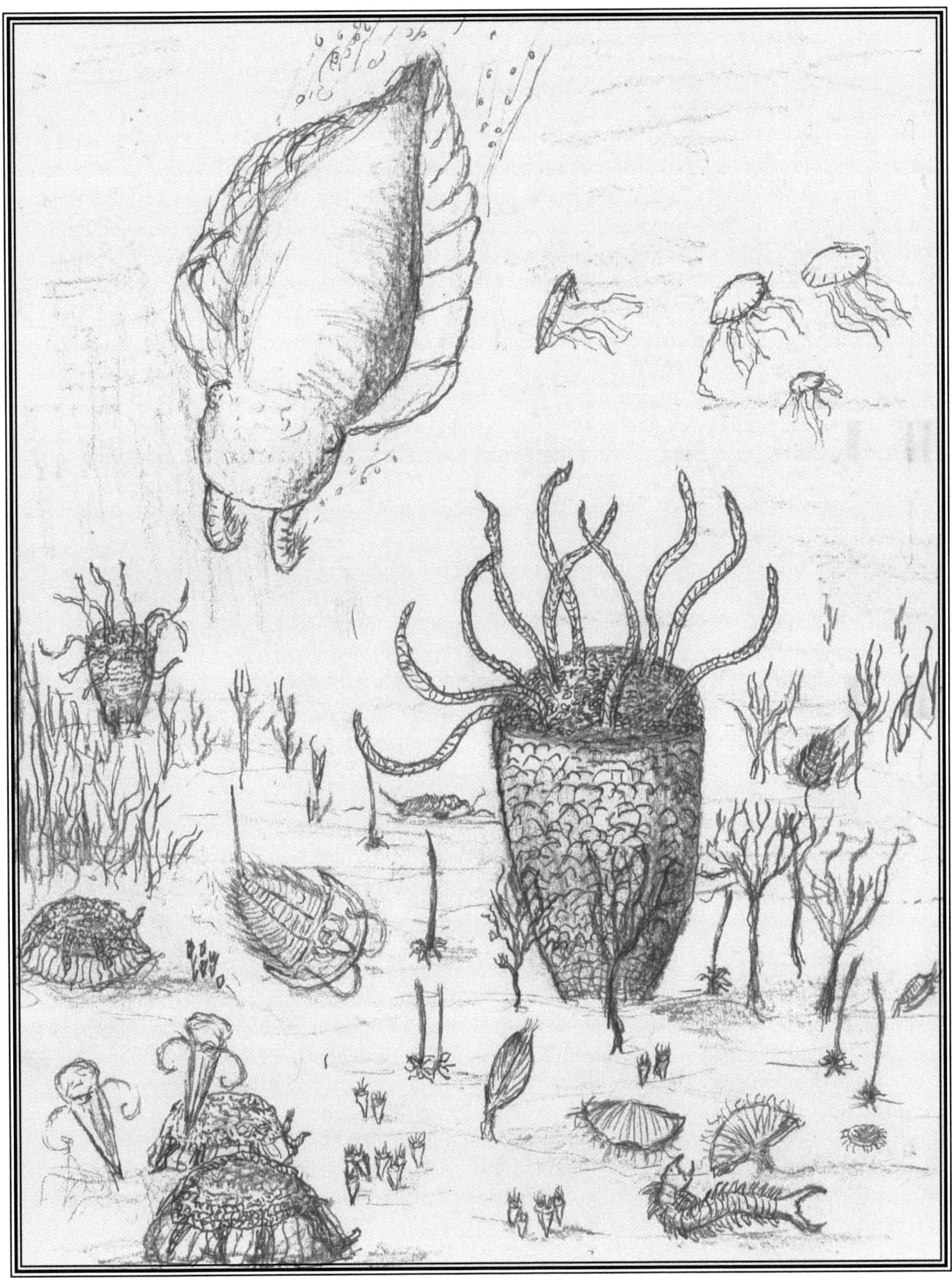

All artwork by: Kerry Matt

Front Cover: ***Pelagiella***
found west of Lititz Pike near Neffsville Pa.

The Lower Cambrian Explosion in my back yard

This is an extension of my first fossil book, **Pennsylvania's Paleozoic_Playground**. I focus this time on detailed field experience in suburban Lancaster Pa. with fauna of the Lower Cambrian Kinzers Shale Formation. The title appropriately necessitates a purpose.

Most knowledgeable collectors and professional paleontologists realize Lower Cambrian time in the fossil record is a less commonly encountered environment. This is primarily based on limited exposures accessible for collecting specimens. We are dealing with a time period much older then most of the fossil record. Many of the remnants of these deposits bearing fossil remains of this early age are simply gone, having been weathered, deformed, or eroded away over hundreds of millions of years. Fortunate to live in such close proximity to an intact paleo-environment offers unlimited access to local sites. This allows one to compile a vast assemblage and variety of fossil finds within a specific area. Fossils from this period are often rare. These are valuable to science, aiding to understanding the nature of evolution of many plant and animal types in the fossil record.

There have been many types of (Pre-Cambrian) fossils in the worldwide record pre-dating our Lower Cambrian fauna (well over 500 million years old). However the use of the expression Lower Cambrian Explosion suggests the idea, when evolution began to explode into a diverse population of life forms. A time when our planet was in its early stages establishing biological stability. Pre-Cambrian fossils for that matter do exist pre-dating the Lower Cambrian. They are comparatively simple with their structural anatomy. Examples include hydrozoa, (jellyfish) fossil remains, stromatolites (algae forms), and strange animals exhibiting coelenterate-like types. Many good specimens of these types rarely crop up in rocks from other parts of the world such as China and Australia. With this book I will do my best to exhibit a portrait of what kinds of specimens were collected locally, what animal they appear to be. Details about the paleo-environment of various sites are included. Some collected specimens may still be incompletely described as to their true nature. A few may be revealed for the first time. Other specimens are likely to represent what others collected in the past however exhibiting supplementary detail in their taxonomy.

It is important to remember the world Famous Burgess Shale's of British Columbia. A plethora of well-preserved fossils from the Cambrian period there were collected and exhibited in museums worldwide. The Burgess shale's are well known for exquisite preservation of soft-bodied animals. This is a bit of good luck for the paleontologist.

The opportunity to reveal so much information from such fine examples allows the paleontologist to examine the process of biological evolution based on exquisitely well-preserved detail. This is almost unimaginable for fossils dating back over 500 million years. My point on mentioning this is to realize our fauna predates the Burgess by many millions of years, an incredibly long period of time. The Cambrian period is subdivided into lower, middle, and upper exhibiting within itself a wide span of development. There are changes in life forms, species, and paleo-geographical status. Specimen preservation within the Kinzers for the most part is not always on par with the Burgess. Nevertheless we too have encountered physical conditions within our formation allowing for soft-bodied animal preservation. There are collected species from the Kinzers Formation that could easily represent proto-types to other known species from the Burgess. This aids in reconstructing the pieces of a puzzle providing a detailed Paleozoic profile.

Trilobite: Olenellus Thompsoni

The abundant and most commonly known fossil found in the Kinzer formation, are trilobites (extinct arthropods). There is group of species having similar anatomical features from the family *Olenellidae*. Though some are extremely rare such as *Lancastria roddyi*. The most commonly known species is *Olenellus Thompsoni*. Disarticulated (broken) specimens of these have been recorded from most of the bedding throughout the entire thickness of a particular, (mostly shale) member within the Kinzer formation known as the *Olenellus* zone. In many places complete specimens are found exhibiting various degrees of preservation. This trilobite was often sub-divided into various species by collectors and paleontologists in the past but to date the singular *Olenellus Thompsoni* is assigned to most of those specimens. Studies of various named species have been re-established over the years. Many were sub-divided on the basis of variables in visible features of surface anatomy. Conclusive evidence proves changes in appearance were often caused by distortion, compression and physical stresses within the rock strata where they were preserved.

Controversial matters with identification applicable to a smaller version of *Ollenellus Thompsoni* deemed *Paedumias* is one example. Later observations portrayed *Paedumias* (appearing very similar to *Olenellus Thompsoni)* as having anatomical variables in comparison primarily based on size. Many seemed to think the so named *Paedumias* was simply a larval or pre-mature *Olenellus Thompsoni*. The species was later named *Paedumias Transitans* meaning this type specimen was in its transitional growth stages to becoming what we recognize as adult *Olenellus Thompsoni*. At present some paleontologists recognizing these trilobites agree certain specimens from collections once classed *Paedumias* were misidentified and they may no longer be considered a separate species. *Olenellus Thompsoni* is the single given name for the species in both its larval and adult form. There clearly appears to be different species from the Kinzers taking on the resemblance of *Olenellus Thompsoni*. Most samples I observed and collected from the field seem to attain a growth of 5 to 6 centimeters in adult specimens.

By comparing specimens of the same size, one can immediately recognize a difference between this species and certain specimens formally proclaimed *Paedumias* (juvenile *Olenellus Thompsoni*). Anatomical features are similar yet obviously different in minute detail. This appears to be notable comparing cephalon borders. Another feature is the narrowness of the thorax. The species appears to be more closely associated with a family of *Olenellidae* assigned to different species such as *Mesonacis* ("old name" *Ollenellus Vermontanus*), *Olenellus Clarki*, or *Olenellus Fremonti*. All specimens in the *Olenellidae* family from the Kinzers bear similar features as follows: (1) a total of 14 segments within the length of the axial lobe including 14 attached plural spines on each side. (2) The third plural spine from either side of the axial lobe is significantly longer than all the others as they taper down along the thorax. (3) At The base of the axial lobe is a long thin telson spine in place of a pygidium. This spine is often as long as the entire thorax, and crainidium together. The *Olenellidae* have no developed pygidium as seen in most typical trilobite species. (4) There is a pair of long genal spines on either side of the cheeks. Smaller, interior, meta-genial spines are usually hidden from view on most specimens. (5) The crainidium exhibits very similar taxonomy but there appears to be subtle differences in *Ollenellus Thompsoni* versus our different species. The cephalic border adjacent to the thorax is tapered at sharper angles on either side of the *Mesonachis*-like species. This feature, genal spines included, looks much thinner on some specimens than as appears on *Olenellus Thompsoni*. (6) The eyes on both species extend to the edge of the cephalon. (7) The glabella usually resembles the shape of a spade including what appears to be a visible stalk extending from the tip of the spade shaped glabella to the border rim of the cephalon. Many specimens from the Kinzers exhibit this feature. Preservation variations allow said feature to be more or less prominent, or at times not visible at all. Could it be possible our trilobite appearing *Mesonachis*-like could actually be the original one named *Paedumias*? Or could this be something new? perhaps an entirely different species?

Basic structural taxonomy showing dorsal view of a typical trilobite from the Kinzer Formation

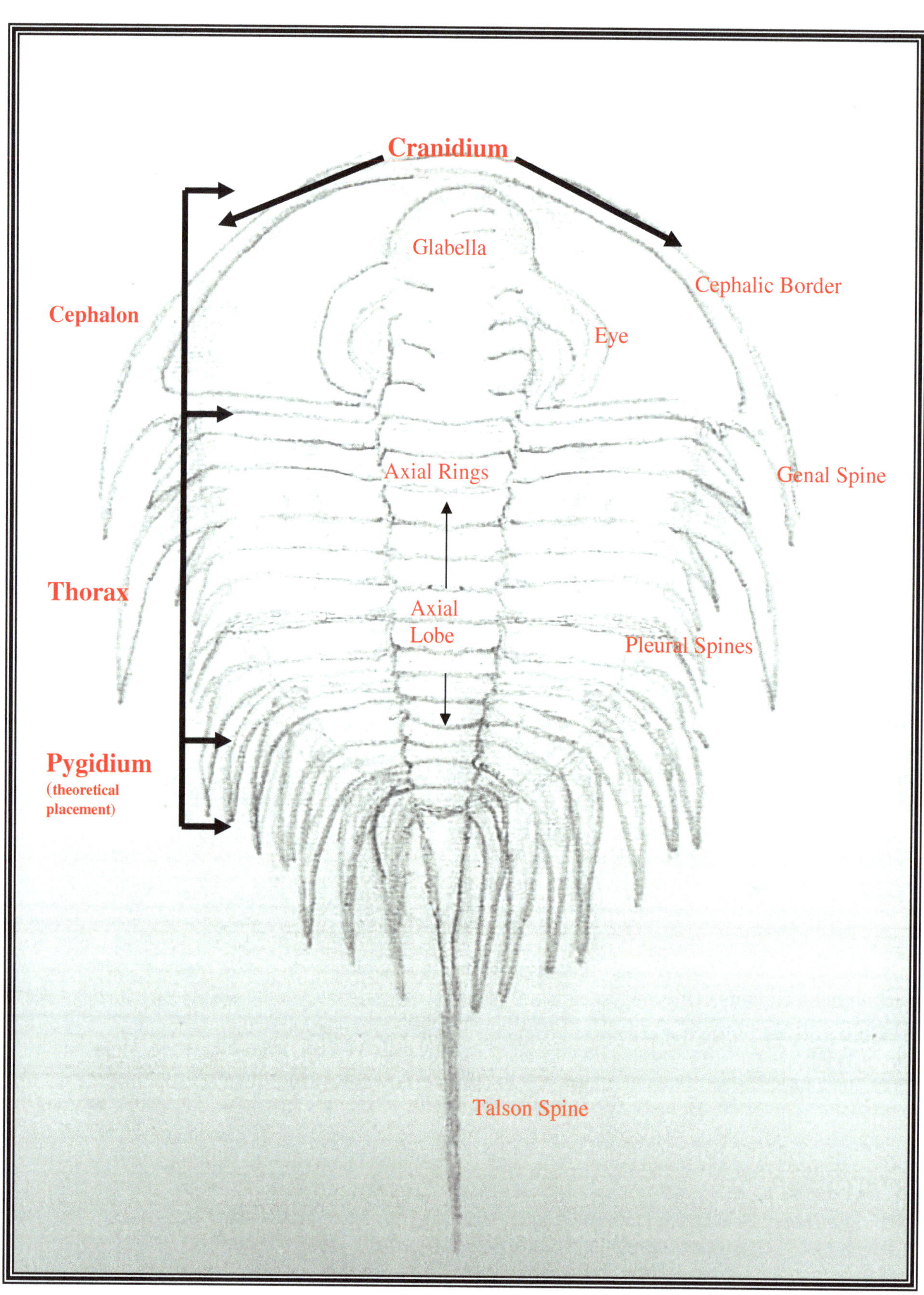

(3)

Trilobite Olenellus Thompsoni

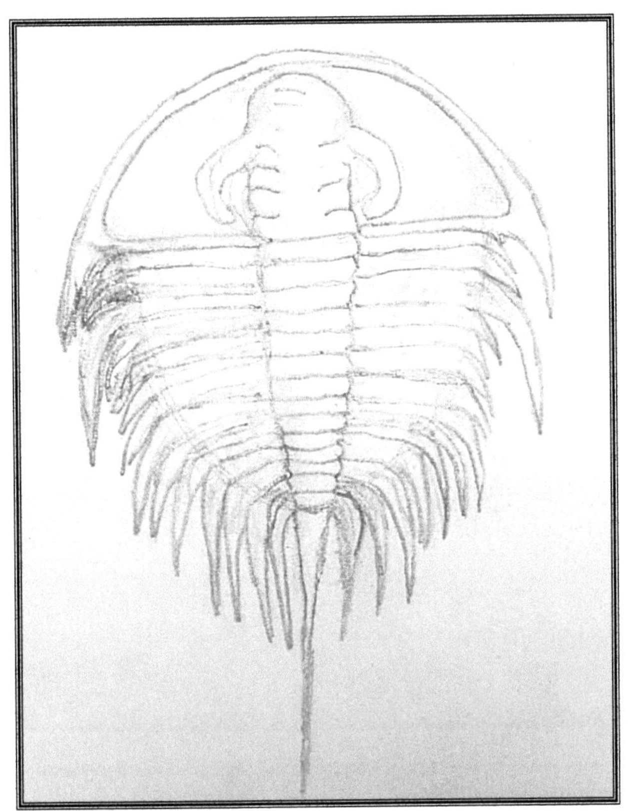

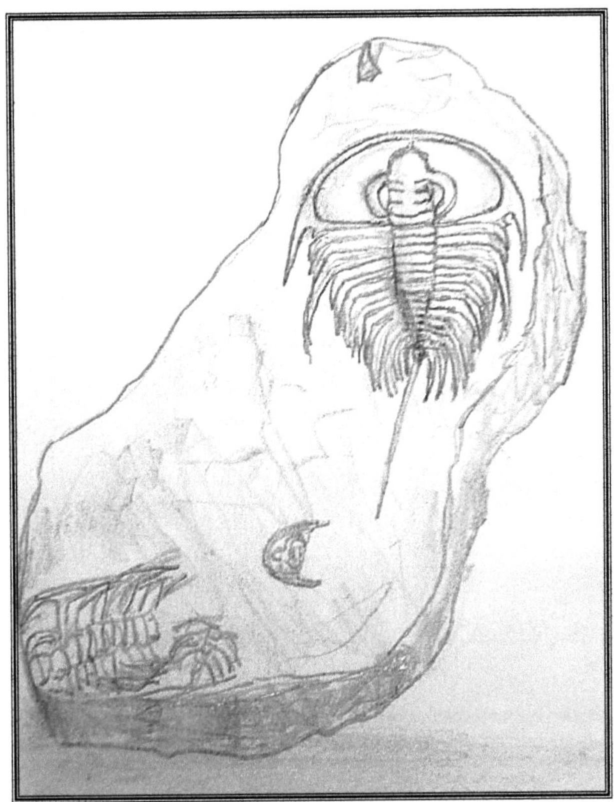

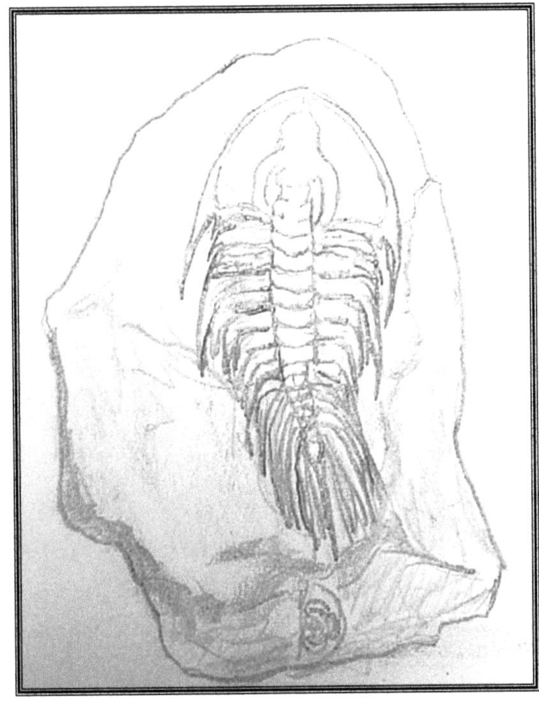

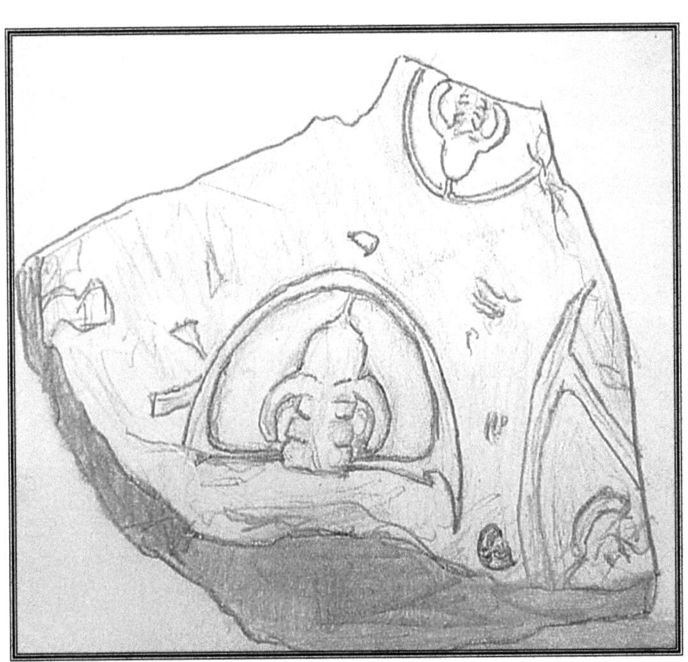

Adult Olenellus Thompsoni

Large head from Getz Woods's 7X14 cm

Getz Wood's 11X16 cm

Adult Olenellus Thompsoni

Both halves of large olenellus from Getz Wood's 12X17 cm

Getz Woods's 8X9 cm

Partial Getz Wood's 10cm

Adult Olenellus Thompsoni

Site off of Harrisburg Pike
5X11 cm

Site off of Harrisburg Pike

Fruitville Quarry 3cm

Fruitville Quarry 3.5cm

Fruitville Quarry

Adult and juvenile Olenellus Thompsoni

Neffsville area west of Lititz Pike +-6cm

Neffsville area west of Lititz Pike +-3cm

Neffsville area west of Lititz Pike

Neffsville area west of Lititz Pike, Juvenile trilobites 5x

Juvenile and larval Ollenellus Thompsoni

Fruitville quarry 30X

Larval stage with Hyolythies
Neffsville area 30X

30x *

Fruitville quarry

3 juvenile specimens, typical of Fruitville Quarry 1cm ea.

* Neffsville area 10x larval stage

Mesonacis-like trilobite

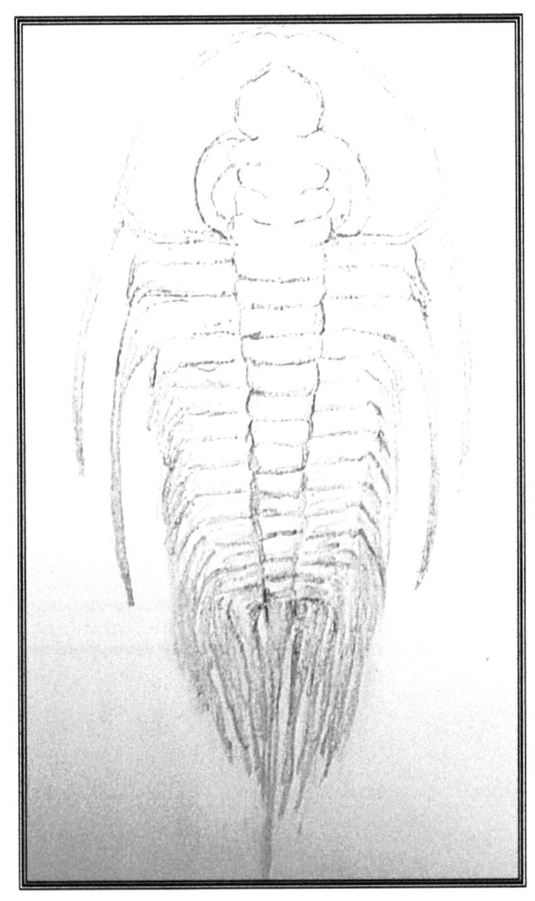

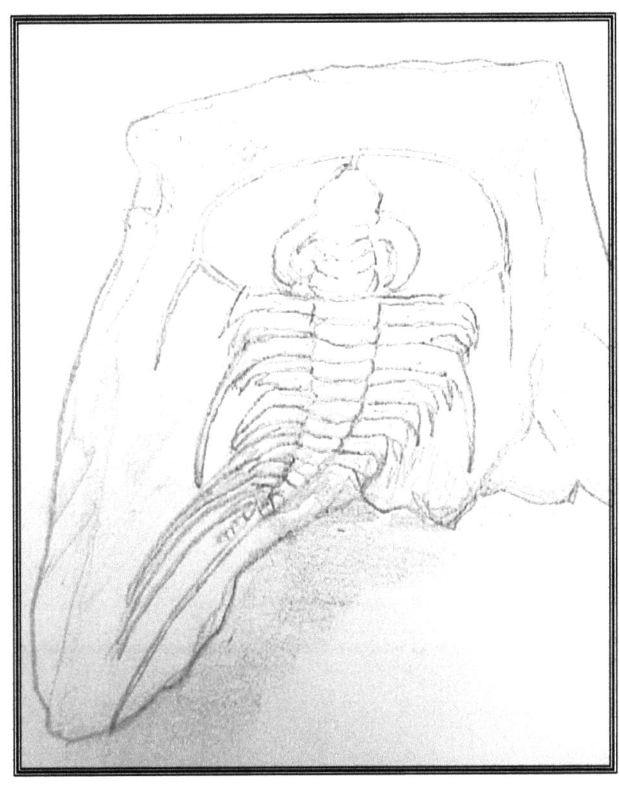

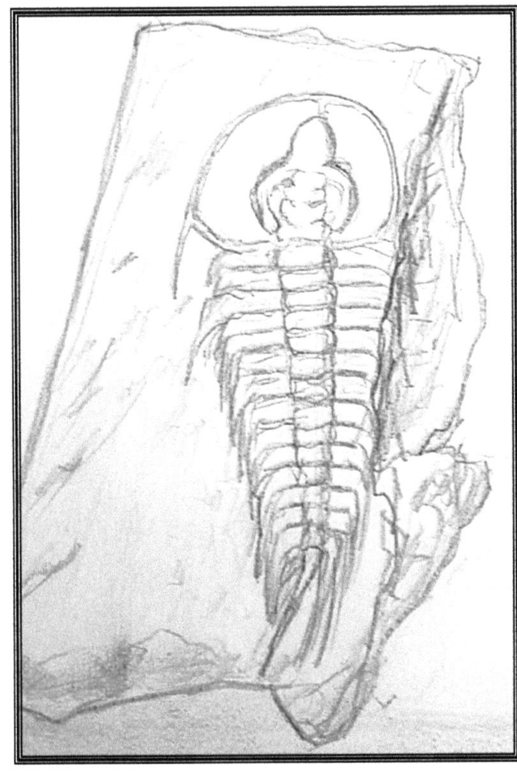

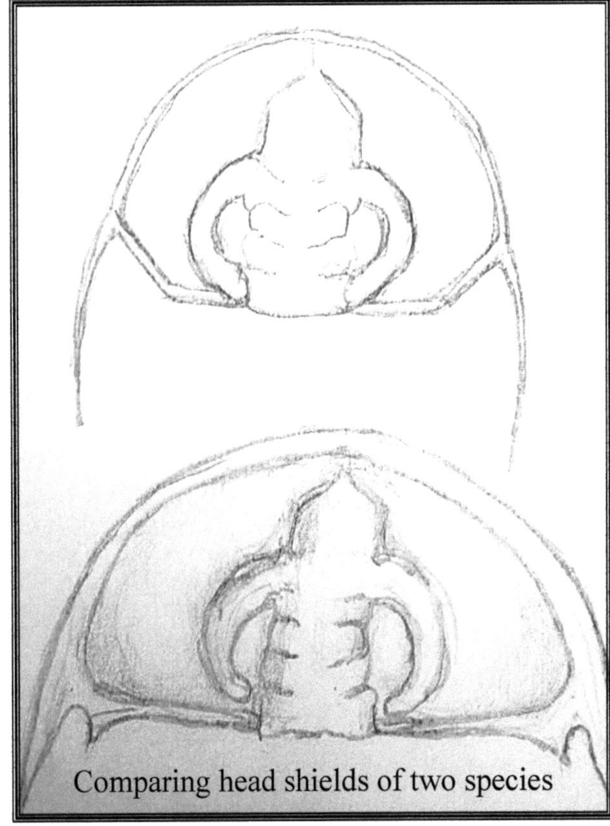

Comparing head shields of two species

Trilobites having Mesonachis-like appearance (site off of Harrisburg Pike)

Mesonacis-like trilobites (site along Harrisburg Pike)

Species from Family Wanneridae

A trilobite very similar to the Olenellidae occurring in fair abundance is the species *Wannearia Walcottana*. This trilobite is considerably rare to most of the world, but prolific in the *Olenellus* zone within the Kinzer formation. It is quite evident the species co-existed with other trilobite species already described. *Wanneria Walcottana* appearing much like *Olenellus* to the common observer has obvious anatomical differences observed by the experienced collector and paleontologist. (1) *Wanneria* exhibit plural spines the length of the body, falling evenly in line along the tips, (2) The third spine is not longer as in *Olenellus*. (3) Instead of having 14 segments both in axial lobe and plural spines the body is longer, having 17 segments. (4) There is no telson spine in the posterior area, but rather a bud like extremity resembling a rose bud or brachiopod in shape. (5) The plural spines extend well beyond the last axial ring and prominently skirt the entire thorax. (6) Another obvious anatomical feature is the distinguishing reticulated skin surface quite often well preserved on specimens collected from the Kinzers. (7) The eyes on *Wanneria* do not extend to the cephalic border along the thorax but are raised significantly above. (8) *Wanneria* has a rather large bulbous glabella with distinguished marginal tapering.

Specimens of *Wanneria* show evidence of becoming quite large. Samples collected in the field show they may have attained lengths of up to a foot long. Articulated specimens of seven inches and larger have been found. Many are maintained in collections and I have collected several excellent specimens. Some of these exhibit cicatrization features and examples such as regeneration of a plural spine. One can see where the tapered end of a plural spine was ripped away showing re-growth. Also damage and re-healing along the cephalon, one example: a specimen showing a possible section where a predatory beast of the period may have bit into the cephalon: another specimen showed a detached hypostoma (feeding part or mouth plate on ventral side) pushed to one side.

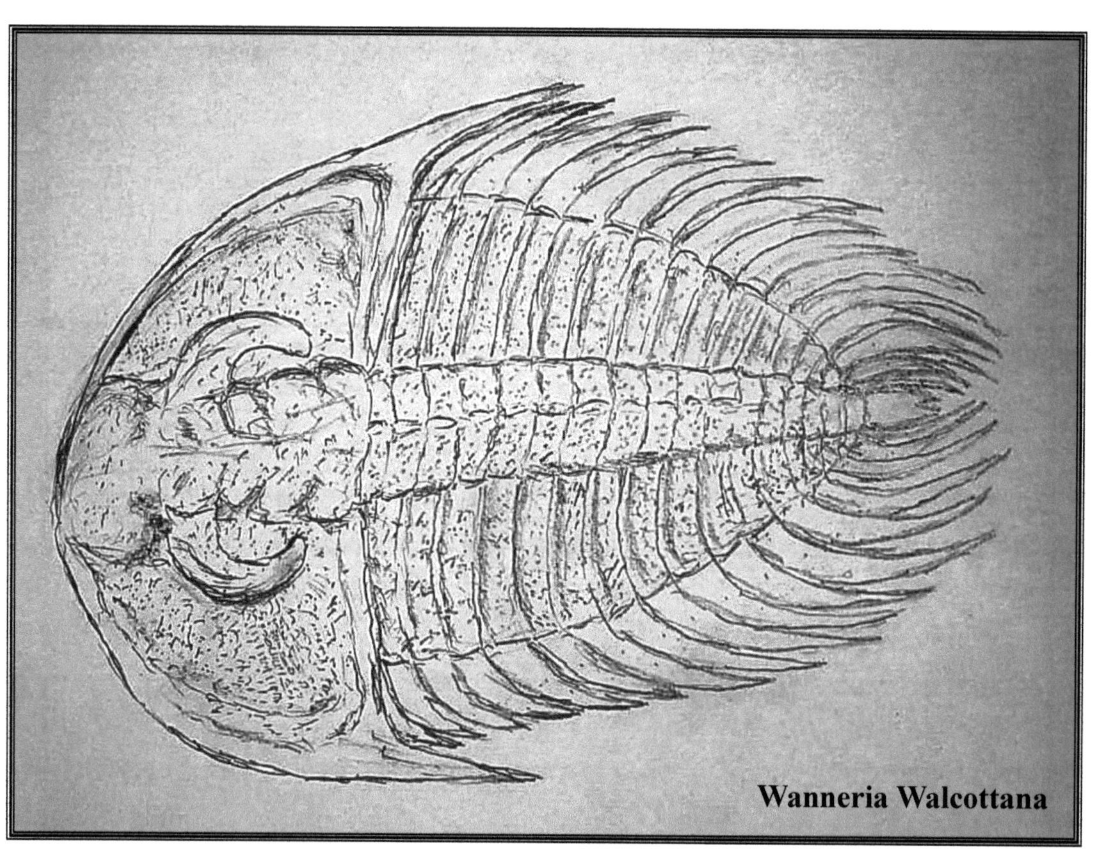

Wanneria Walcottana

Wanneria Walcottana
World Class Examples from Harrisburg Pike area

Wanneria Walcottana
(examples from Harrisburg Pike area)

Showing Hypostoma and plural spine regeneration*

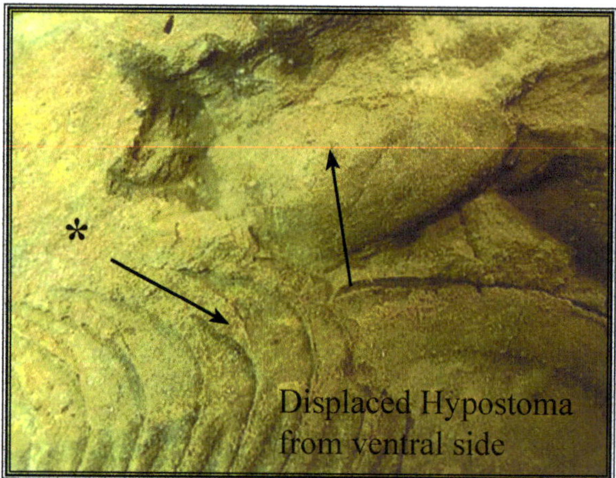

Displaced Hypostoma from ventral side

+-15 cm

+-11 cm

Note* pathological destruction

+-7 cm

+-11 cm

Wanneria Showing detail in skin taxonomy (Note* Predator bite along cephalic border)

30X Note hex-like skin pattern

Skin pattern in Pleural section

Pattern on interior of a cephalic border

Wanneria Walcottana

Pair of Wanneria Walcottana from Long Farm Site near Neffsville (construction site fall 2006)

Painted and non-painted casts of examples collected from Fruitville quarry

Wanneria Walcottana 13.5cm (7”)
from Harrisburg Pike area

Trilobites from the Bonnia family

Earlier fossil collectors and paleontologists reported various rare specimens from the Kinzers lower member within the *Ollenellus* zone, particularly from York County west of the Susquehanna River. Spectacular specimens of *Ollenidae*, and *Wanneria*, were found through out the formation in York Co. The trilobite *Bonnia* and related types primarily occur in sandy or coarse-grained bedding lying directly on top of the Vintage Dolomite. Often found together with brachiopods, these were mostly disarticulated remains of small trilobites with taxonomy unlike the *Ollenidae*. Specimens were mainly small heads, having simple eye lobes, glabella, and pygidium like those found on more commonly observed trilobites from later groups. Several species associated with *Bonnia* have similar taxonomy. Personal finds of specimens have not developed enough sufficient variables to differentiate species from similar litho faces emerging at sites that I worked in Lancaster

My first encounter with *Bonnia* trilobites was in a site along Harrisburg Pike. Sandy bedding overlying the Vintage Dolomite yielded an abundance of heads and pygidia. These often occurred with *Olenellus Thompsoni* (usually disarticulated parts) and a very small scalloped brachiopod. Other collectors have brought a few fine examples of articulated *Bonnia* to light. I was fortunate to find a few nearly complete specimens. For those familiar with the popular Ordovician trinucleidae trilobite *Cryptolithus*, I suggest head shields of *Bonnia* resemble the cranidia. The thorax and pygidium appear more like the common Devonian *Phacops* or similar types. I wandered if this horizon in the litho-face could represent a near shore deposit. The presence of brachiopods and a mostly disarticulated yield of trilobite remains may have been caused by agitation (waves, currents etc.) within this stratigraphic zone.

The same apparent bedding was documented at a construction site on a tract known as the Long Farm. The bedding there seems to be much thicker by a great margin though I found no brachiopods at the site. In contrast there was a high population of a very small conical mold known as *salterella*. Another construction site near Neffsville, a few miles due southwest of the Long Farm site northeast of Fruitville quarry, exhibits apparent disconformity. There are similar characteristics with bi-valve fauna relating to the *Bonnia* layer. The material was scattered as float lying above the Vintage Dolomite in a mass of clay with sinkholes present in the vicinity. It seemed this bedding slumped away from the top of the underlying Vintage Dolomite resting in contact with Antetum sandstone (also scattered in the clay) perhaps due to faulting. The material found here was of very special interest because of the good preservation of soft-bodied animals, and abundant brachiopods, which will be discussed further in the text.

Within isolated fragments collected were examples of disarticulated trilobites similar to *bonnia* but from other species I was not familiar with. There was one partial trilobite called *Olenoides*, which I never observed elsewhere in the Kinzers. None of the *Bonnia* types or otherwise appeared to be present in the upper member of the *Olenellus* zone adjacent to the overlying Ledger Dolomite. There are high populations of *Olenellus* and *Wanneria* having abundant yield in juvenile stages in much of the Kinzers upper member. It appeared as though remaining rock found at this site (between Long Farm and Fruitville quarry) originated from the lower member of the *Olenellus* zone composed of much finer sediment than usual. I am imagining a shallow marine mud flat with a different mode of preservation than the typical sandy strata. This was a seemingly rich paleo-biosphere that received plenty of sunlight. The presence of algae-like fossil remains, coelenterate-like creatures, and shallow marine brachiopods from material collected contribute to bringing forth this postulate. I will attempt to further summarize this as we continue a detailed observation of the remaining fauna.

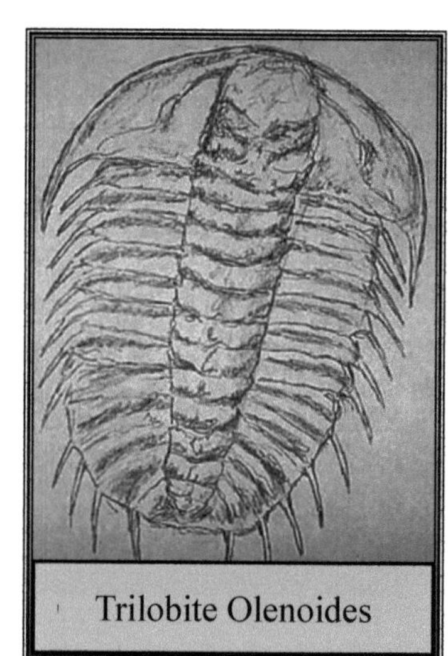

Trilobite Olenoides

Trilobite Bonnia from site along Harrisburg Pike

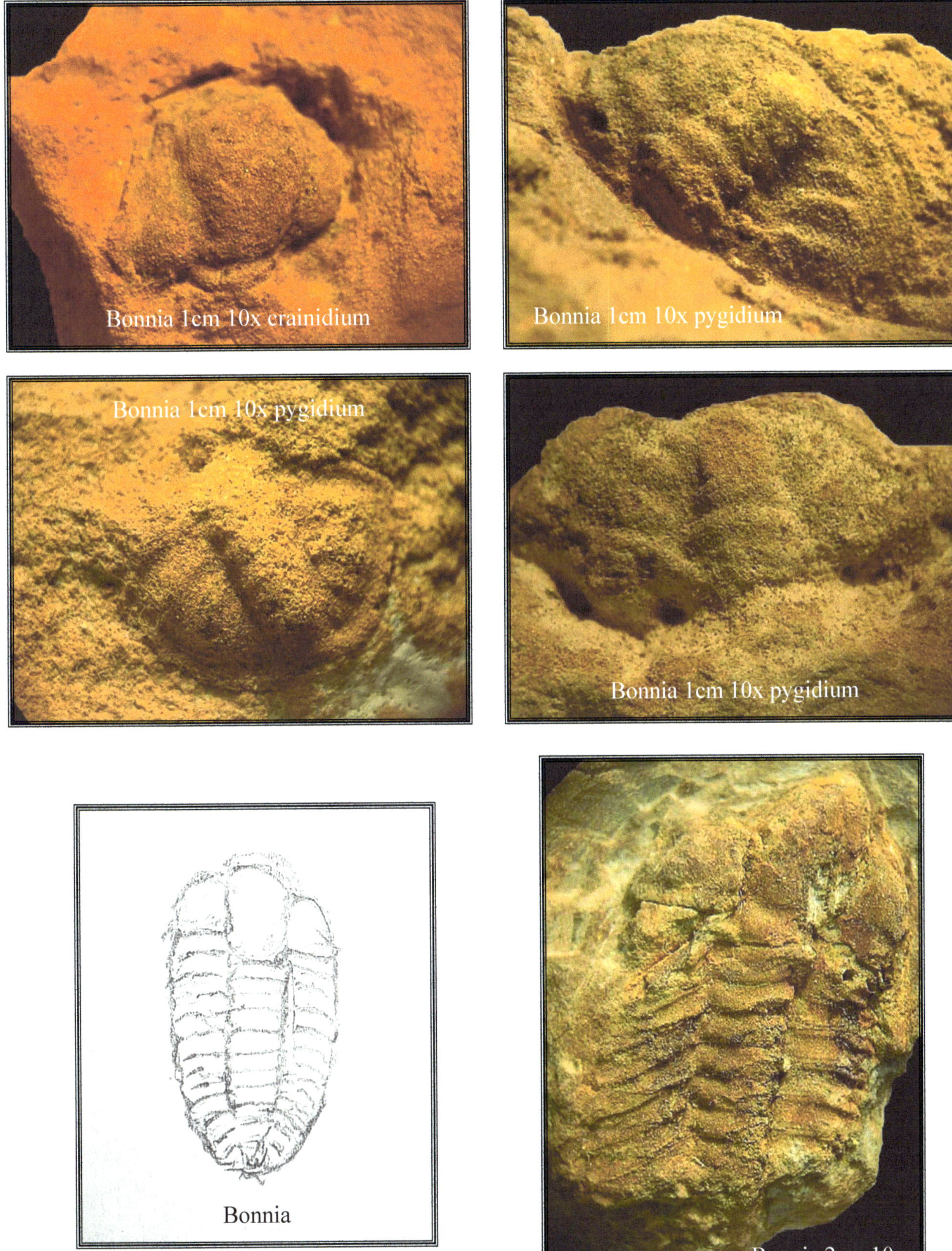

Rare Trilobites: construction site near Neffsville

Un-identified Cephalon with brachiopod 10x

Partial Olenoides 1cm 10X

Un-identified Pygidium 1.75cm 10x

Un-identified cephalon 1.50 cm 10x

Pygidium 10x .75cm possible trilobite Kootenia

Bonnia-like head with surface detail 1.25cm 10x

Rare Lower Cambrian Arthropods

One of the most unusual significant finds from the local vicinity was from a rock out-crop in the Kinzers formation once quarried for building stone in the Getz woods area. Here the formation takes on a different texture than most Kinzer rock from the surrounding area. The shale was dark gray with perhaps a tint of bluish tone mixed in. The clastic nature of the bedding was somewhat a porcelain-like texture breaking in rather sharp shards. The bedding and cleavage were both a challenge to work with, most times parting at close angles. The body of rock was distinctly within the *Olenellus* zone. Many specimens were recorded from the exposure. For a novice collector it was not uncommon to break rock along the cleavage to find dendrite deposits of an iron oxide appearing as if to be plant-like remains. Within the true bedding this often gave fossil remains a signature patina of vivid reds, yellows, and orange. Fossils from the site were usually well preserved in this fine grain hard shale. Institutions within the commonwealth still harbor many specimens of large trilobites

From my experience I was able to locate a bedding plane that was very productive. The majority of the specimens from this site were large adult *Olenellus Thompsoni*. Juvenile trilobites also appeared but not in great abundance. I found little evidence of larval specimens. It seemed *Wanneria Walcottana* were not abundant. I found no specimens at all from that species until different bedding was excavated a few hundred feet from the original location. I found a complete *Wanneria* in a very pale gray (nearly white) chalky rock, which seemed to be an isolated specimen. In my opinion this deposit was perhaps a shoal where deeper water only allowed finer sediments to accumulate producing the smooth fine-grained texture. I pictured deep water with mainly large trilobites. There may have been colonies of echinoderm like creatures on the floor such as *Lepidocysti*s. There seemed to be no brachiopods in that part of the formation.

Of special interest was my find not appearing to be the usual trilobite. I was breaking shale, then noticed plural spines oddly arranged from the way they should normally appear. At the time what caught my eye was barely showing. I was almost going to toss the specimen as a "blown out" trilobite. I thought I better inspect this again more carefully and get it back to the lab. The bedding indicated a direct split for a part-counterpart would be impossible. Rather than take the risk I proceeded to flake the rock away to produce a solitary specimen of what ever might appear. I was very impressed with the anatomical features slowly appearing before my eyes. This was no trilobite at all, but something quite unusual. A thorax was emerging reminding me of a large shrimp-like body, only having four segments with tapered pluron like parts but no axial lobe. Then the real surprise: The tail exhibited a pair of long spikes with minor spines next to each. The head was featureless but seemed to be fully intact. There also appears to a spine-like feature, which was probably a superimposed set or pair along the body. The head seems to be shaped more like that of a trilobite, appearing as if it were folded back over the thorax at an angle.

This beast immediately struck me to be one of the possible enigmas leaving evidence of predator damage found on trilobite remains throughout the Kinzers. My guess was this was an arthropod classed with phyllocarids. Perhaps this was similar to the likes of *Anomalocaris* (thought to be a carnivorous predator). It may be difficult to tell weather the specimen is exhibiting the dorsal or ventral view. I have not encountered any other examples in exhibits, literature, or in the field since this find. The specimen has superficial taxonomy in comparison to a Burgess species called *Canadaspis*. The truth of paleontology with respect to this demonstrates a cutting edge to this science. This might be a discovery of a new species, or possibly a proto type. Perhaps finds like this can provide additional information needed to connect pieces of a pre-existing puzzle.

Rare arthropod from Getz Woods

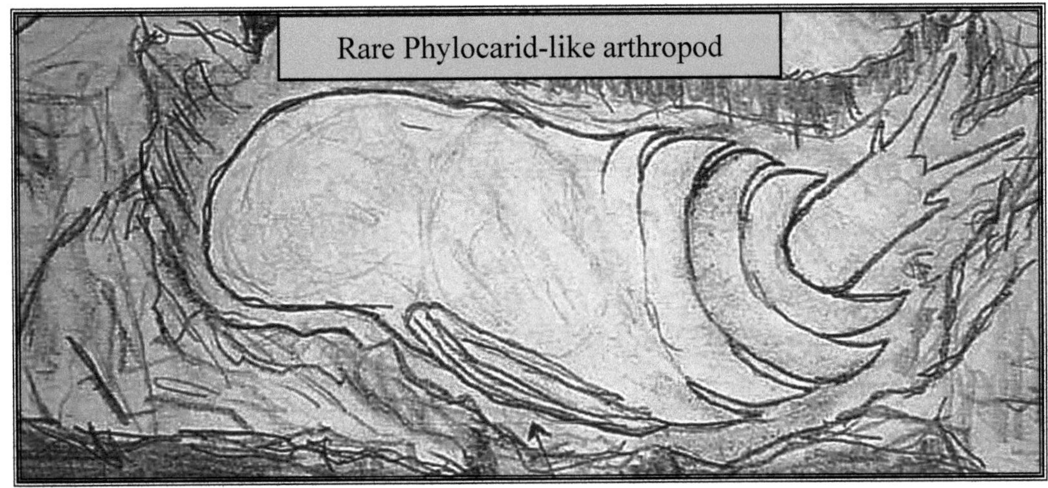

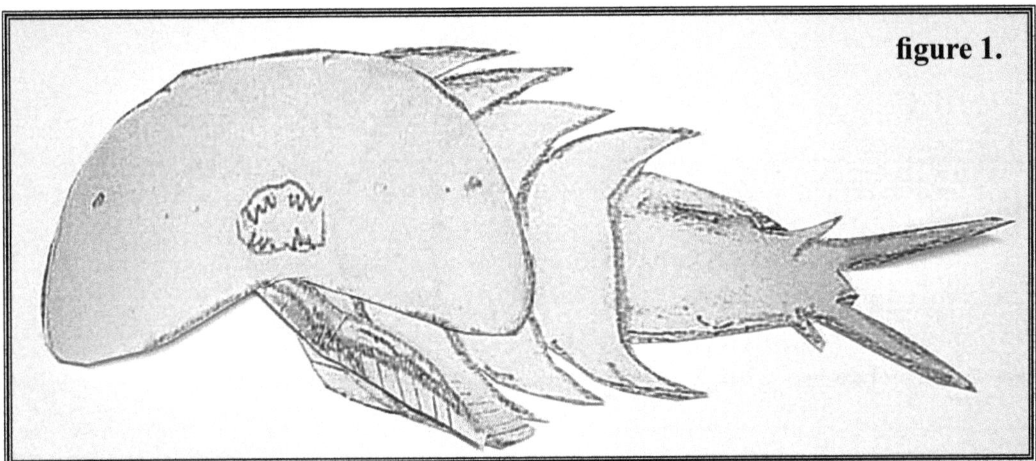
figure 1.

In theory it appears the head **figure 1**. was twisted around and superimposed on top of the thorax. A centrally located set of either swimming spines or mandibles may have protruded from a common central location folding to one side becoming superimposed along the trunk of body. The model (drawn with transparent head displays pseudo mouth and eyes showing through) it could depict either side since there are no visible signs of a mouth or eyes on the actual specimen. The head could in reality be the ventral side entirely flipped over on the dorsal side of the body **figure 2.** Represents a hypothetical model showing the ventral side of the animal in its normal structure in life based on an anatomical theory.

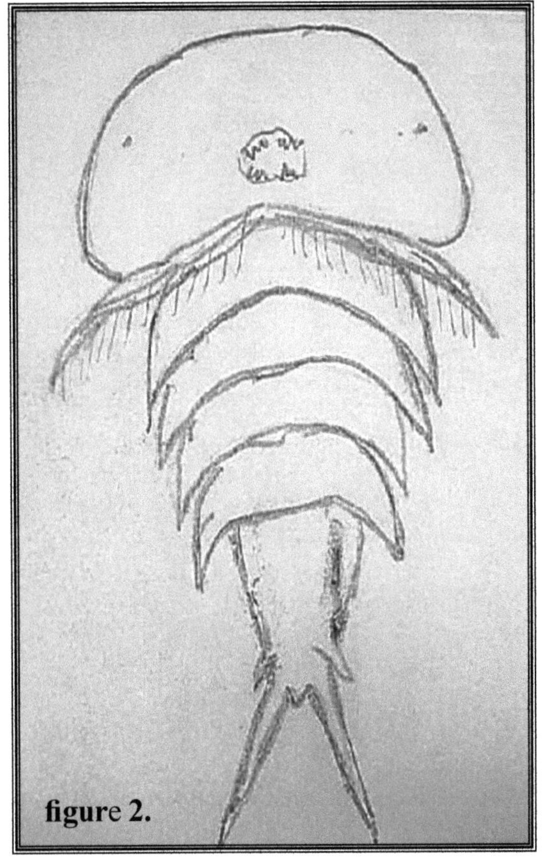
figure 2.

Rare Arthropod, Phyllocarid

New construction sites from areas near Neffsville produced a spectacular array of fossil specimens from many classes. Not only were there commonly known trilobites but also abundant mollusk specimens of perceivably rare and uncommon types. A light tan colored soft-grained shale occasionally produced excellent preservation of a few soft-bodied animals, as well as aquatic plant remains. Most of the soft-bodied remains were not as fine detailed as the Burgess Shale, or magnificent finds from China. They were enough to put the local fauna on the map for closer study

Specimens I gathered represent unusual types of trilobites, crustacean like arthropods, brachiopodia, a plecyopod, annelids, sponges, possible coelenterates, and algae-like plant forms. The difficult part of collecting complete specimens had a lot to do with the strata. The rock at the locality was distributed in a clay bed. No litho-face was found intact inhibiting a better yield. Pieces were gathered mostly near the surface. These were split apart producing the finds listed in this text. On one occasion, I found what appears to be a phyllocarid. In fact, the plate of rock could have produced multiple specimens, if it were not broken through one of them. The evidence was clear enough to show a repetitive organic signature of a living organism suggesting this classification. It appeared as a complete carapace exhibiting a possible eye, with thinner body emerging showing talson areas resembling a dauphin tail. The partial (tail only) was an identical match. I considered this to be repetitive anatomical evidence concluding a definitive duplicate of the other specimen on the same slab. The specimen could be on a comparison to a species known as *Odaraia*, found in the Burgess shale of British Columbia.

Rare Arthropod Phyllocarid

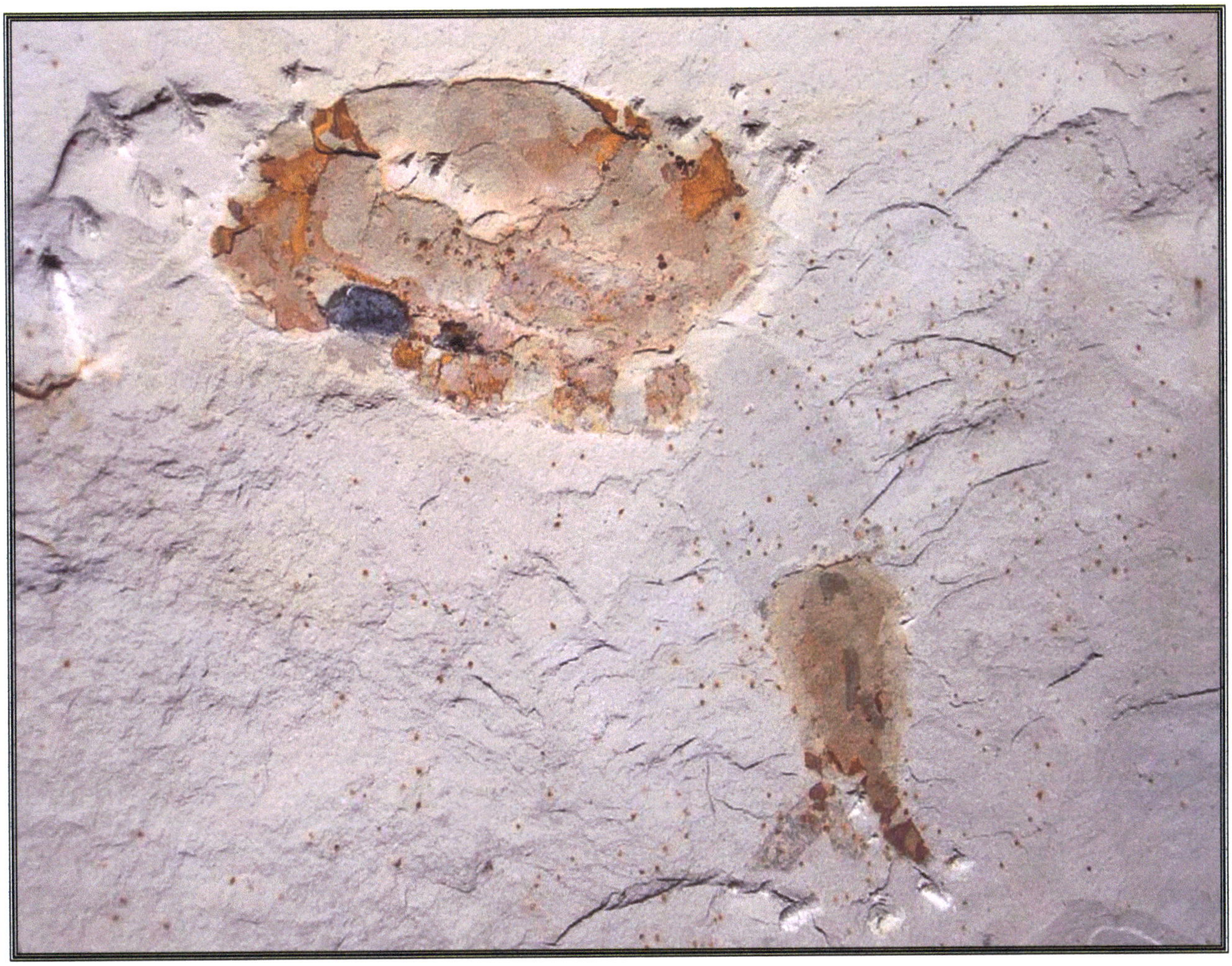

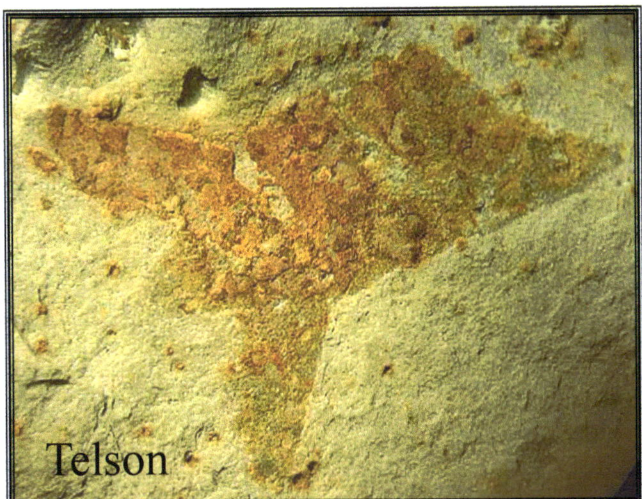

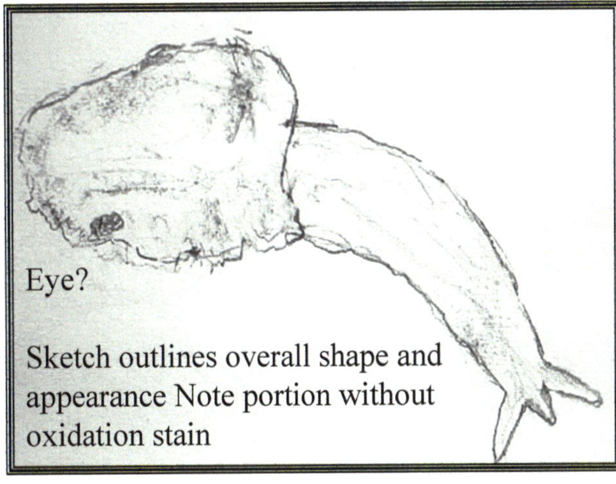

Telson

Eye?
Sketch outlines overall shape and appearance Note portion without oxidation stain

Arthropod Serracaris Lineata

Another controversial species found in the Kinzer formation was recorded in the past by paleontologists. In reports, various finds of fragmented specimens showed up as filamentary carbon details on Kinzer shale. Fossils exhibited what appeared as a long segmented body with short pointed plural extremities. The name *Serracaris* was given to the specimen because of the saw blade-like appearance. I have encountered specimens representing this species in pale fine grain rocks exposed at the Neffsville site. A few specimens collected exhibited portions of a narrow segmented body with several plural attachments intact. I also found a large portion of a telson showing detailed structural outline. The fragments were mostly oxidized with a thin coating of brownish iron residue. One unusual specimen was possible suspect of being a head to this species. To the best of my knowledge, no completely articulated specimens were ever recovered. The complete anatomical appearance may still be unknown.

My illustrated attachment of the so-called head of this animal is hypothetical. There are many features contributing to this being a possible part of the model. One being the specimen was found in relatively close proximity to the other remains (though not at the same time). The size and shape seem to fit the picture. The model is simply meant to suggest how a complete specimen might appear if indeed this fossil is part of the taxonomy. The specimen strongly suggests having mandibles on either side of its bi-symmetrical structure. The piercing parts or tips appear in an oxide residue suggesting a different mode in the decay and preservation, obviously due to the material being harder than the rest of the body. One feature of importance is a filament of dark material on the right boarder of the apparent cheek. It shows a distinct thorn-like rim along the border with intermittent spines tapered along the crest of the cheek. Unfortunately the opposing side does not exhibit the same pattern. One can only presume the pattern is folded or lost. The nature of this feature also suggests a saw like appearance. This specimen is suggestive of affinities to the overall taxonomy of *Serracaris*. It is difficult to tell weather there are eyes showing beyond a few indefinite features. They may be physical imperfections on the specimen. The V shape of the central portion between the so-called mandibles suggests the concept of aiding to funneling or trapping food. My sketch model is based on a personal theory as to what articulated *Serracaris* might have looked like. The model is not necessarily declared to be one hundred percent scientifically accurate.

Partial body and spines of Serracaris

Arthropod Serracaris Lineata

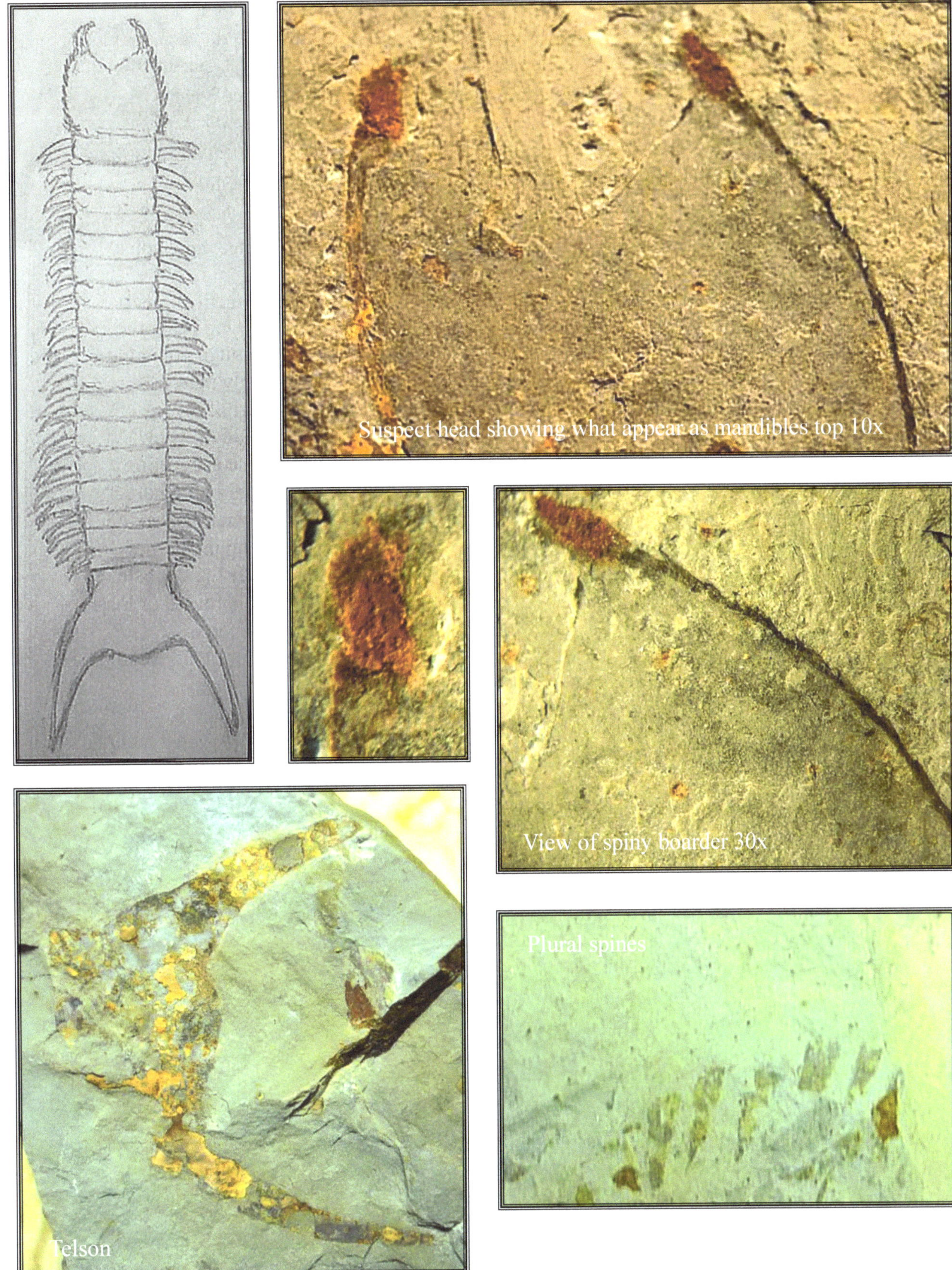

Anomalocaris and other unusual arthropod-like specimens

It always seemed likely to those of us collecting in the field there was a predator out there amongst our quarry. The exact nature of such a creature was an enigma. We occasionally observed specimens exhibiting evidence of being attacked by an elusive soft-bodied creature. Many we observed occurred in a rock not conducive to preservation of such creatures. For all of my personal field experience it was not until the exploration of the Neffsville area that potential evidence came to light.

My first specimen came as a small feeding appendage known to be part of Anomalocaris. This was similar to those collected from the Burgess shale of British Columbia. My specimen was a considerably rare find. It was only a few centimeters in length. Unfortunately I no longer have the sample or any photos of it to share. Later I was lucky enough to find another larger specimen. This time it had detailed preservation unsurpassed by any other found in the Kinzers. Originally the feeding apparatus of *anomalocaris* was thought to be a separate animal in its entirety. It had a nature seemingly independent of being part of a much larger animal. As for previously collected specimens, their affinity wasn't very well understood.

This represents a classic example of how paleontology remains a cutting edge science. Pieces of a puzzle later become clearer as new data is collected in the field. The latest sample I collected was preserved in pale fine-grained shale. The remains appear to be a carbon filament showing exquisite detail. The length of the specimen is segmented almost having a three-dimensional appearance over all. The specimen clearly shows a pair of laterally arranged rows of spines. This was believed to be one of a pair of grasping appendages of *Anomalocaris* used to clasp while holding its prey, then devouring with a circular set of toothed jaw parts. I never found jaw-parts in the field within the Kinzers.

Another peculiar specimen appeared to be paired layers of possible remains of a small anomalocaris body. The specimen shows a form of an oxidized pattern on a rock resembling the body outline. The specimen is not well preserved enough to say for sure. It appears to be dorsal and ventral with filled space inbetween the collapsed or flattened body. There are features suggestive of being more like another species from the Burgess shale called *wiwaxia*. Vague overlapping structures appear on the surface. The *wiwaxia* theory for the specimen is in-definitive and pure speculation.

Anomalocaris and other unusual arthropod-like specimens

Could this be the body of Anomalocaris or perhaps something like the Burgess Wiwaxia?

Could this be Emmonaspis Cameensis Walcott?

30x

Small-unidentified specimen 30x

Segments and gut undetermined species

Unusual arthropod-like specimens

Could this be the surface of a carapace like that of Sidneyia, Emraldella or something similar from the Burgess?

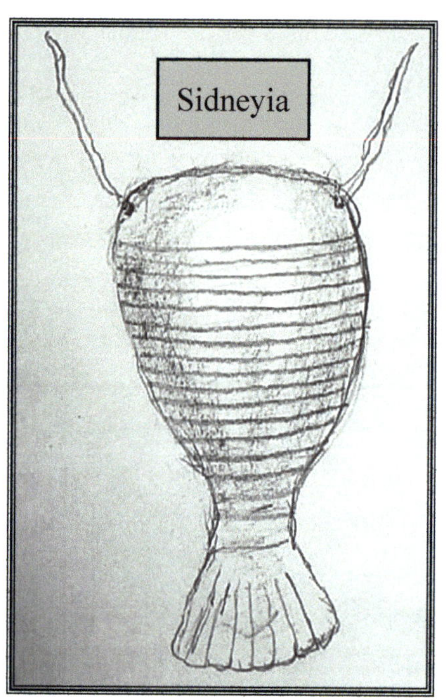

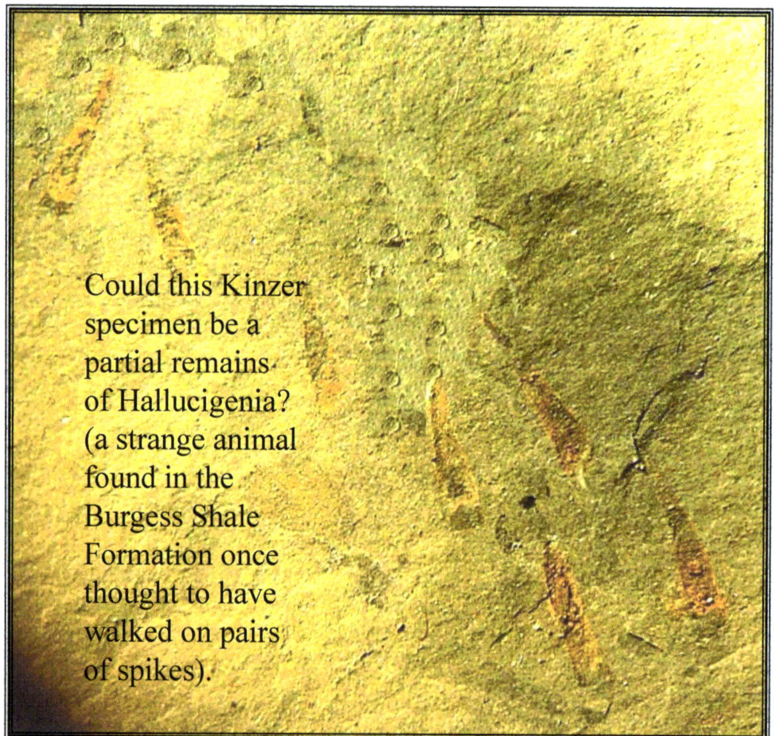

Could this Kinzer specimen be a partial remains of Hallucigenia? (a strange animal found in the Burgess Shale Formation once thought to have walked on pairs of spikes).

Mysterious fossil remains

Echinoderms From the Kinzer Formation

In the mid 1980's, while working a site along Harrisburg Pike rich in trilobites there appeared strange oval shaped blobs in the shale beds. Close examination revealed detailed organic structures within these peculiar forms usually ranging an inch or two in diameter. I soon became familiar with the earliest type echinoderm known as *Camptostroma Roddyi*. One might imagine this to be a prototype for crinoids, sea cucumbers, sea apples, starfish, urchins, and sand dollars. These organisms flourished together with trilobites within the *Olenellus* zone. Many specimens were collected. The best ones exhibited structural details displaying a ribbed appearance. This animal exhibited a platy skin surface including a few tiny internal structures of pitted circles with fringes resembling a sunflower. The true nature of *Camptostroma* seems difficult to model or reproduce. I always figured it to be somewhat like a cystoid in nature. In life, it probably appeared lying on the sea floor much like a modern sea cucumber. I found one example exhibiting tentacle-like extremities protruding from the presumed face of a specimen showing a reticulated pattern within the arm structure

Another Harrisburg Pike site emerging near Landisville later produced many more specimens. One peculiar enigma was a cluster of *Camptostroma* in which the individual specimens seemed to take on a pentagon shape while attached together. This peculiar specimen brought questions to mind. Were the individual *Camptostroma* we generally see just fragments of one much larger creature? Did individuals often huddle together in colonies? I only ever found one with tentacles and I imagined matted groups of the *Camptostroma* where only certain ones had tentacle parts. Could it be one very large animal with many similar interior divisions and certain sections along the outer parameters bearing tentacle-like extremities?

With this observation I am not attempting to declare the final answer as to just what *Camptostroma* was like. I am expanding hypothetical thoughts based on finds from the field. I still think of how the feeding parts of *anomalocaris* were once assumed to be separate creatures of their own kind. They later turned out to be only portions of a larger disarticulated animal. Could the true *Camptostroma* have been a complex animal easily disarticulating after life to become scattered about the sea floor? At the time of this writing many specimens are surfacing from a construction site east of Lititz Pike. One example of *Camptostroma Roddyi* found in a hard gray rock exhibits excellent radial symmetry and grooved rays. From this site it appears the species was common through most members of the Kinzers from lower to upper. Many specimens were also collected from the Fruitville quarry but I did not find any there myself.

Camptostroma Roddyi displaying extremities Harrisburg Pike Site.

Echinoderms from the Kinzer Formation

Echinoderms from the Kinzer Formation

Structural details of Lower Cambrian echinoderm from Kinzer Formation Harrisburg Pike

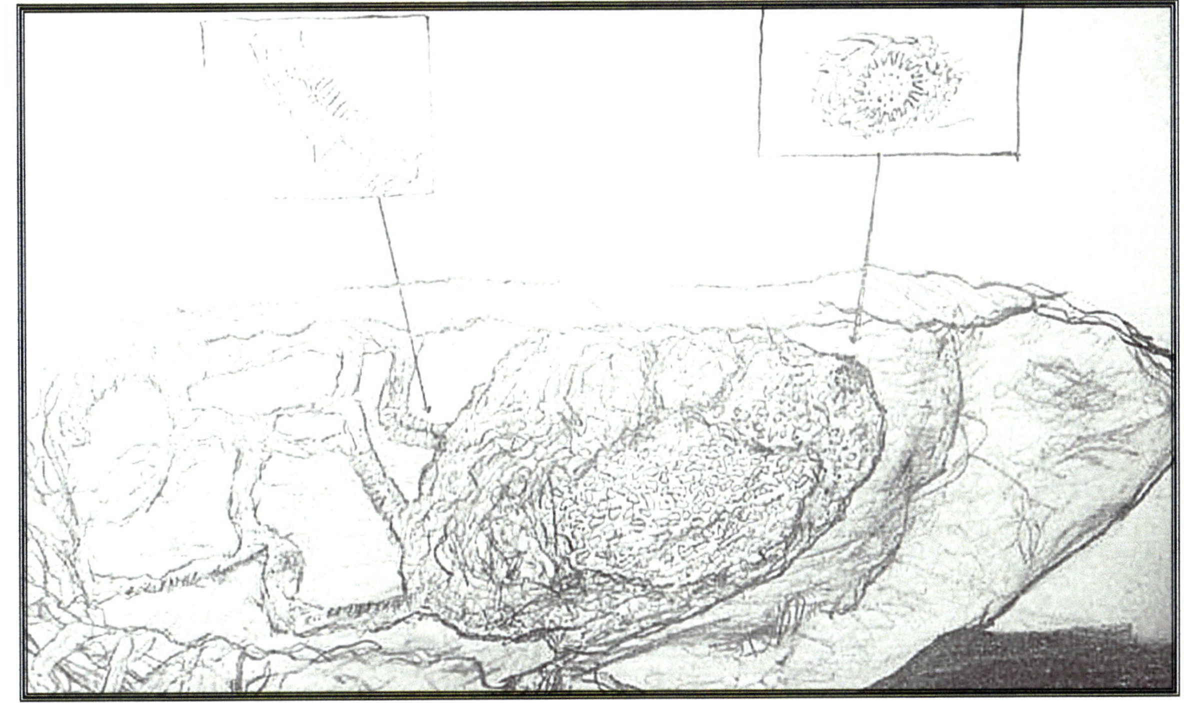

Echinoderms from the Kinzer Formation

Cluster group from Landisville site

Multiple Camptostroma Landisville site

Echinoderms from the Kinzer Formation

Excellently preserved Camptostroma Roddyi from site east of Lititz Pike near Neffsville negative top - positive bottom

(36)

Echinoderms from the Kinzer Formation

Another echinoderm species found in the Kinzers formation are *Lepidocystis wanneri*. Disarticulated plates seem to be found matted along bedding planes of various fossiliferous beds. It may even be possible some of the *Camptostroma* samples described earlier in the text may actually be *Lepidocystis,* particularly those bearing tentacle-like extremities. It seems the two species have similar characteristics.

One specimen in particular that I found at the Harrisburg Pike site has an unusual but obvious difference in taxonomy. It was one of those specimens like the Getz Woods arthropod where I glimpsed something appearing unusual on a rock that I was about to toss away but could not let go. Another look and I knew I should take the specimen home, clean the mud off and get a better view. This appeared to be an echinoderm for sure. It had me thinking, reminding me of an Edrioasteroid. It was small in size, around a half of an inch appearing as a spherical or bulbous mass with superimposed structural detail on the surface. Close observation under the microscope, tweaking my visual skills allowed me to sketch this unusual fossil for aid in study. It appeared three out of five arms were visible around the circumference. Repetitive surface details seemed to represent a structure like that of Edrioasteroid. Three visible arms appeared to be separated at 72-degree angles (five ray symmetry) allowing for the presence of two more possibly imbedded behind the exposed part of the fossil. I was able to find Edrioasteroid could exist into the Cambrian period. I wonder if it could be a new species of the kind? It was suggested this could also be a small portion of the detailed central part of a *Lepidocystis*. Perhaps it is an exposed mouthpart with visible feeding apparatus in tact.

(37)

Echinoderms from the Kinzer Formation

Arrows point to three arms surrounding the spherical body 10x

Drawings help to emphasize subtle details in taxonomy of this Edrioasteroid-like echinoderm found at the Harrisburg Pike Site from the Kinzer Formation

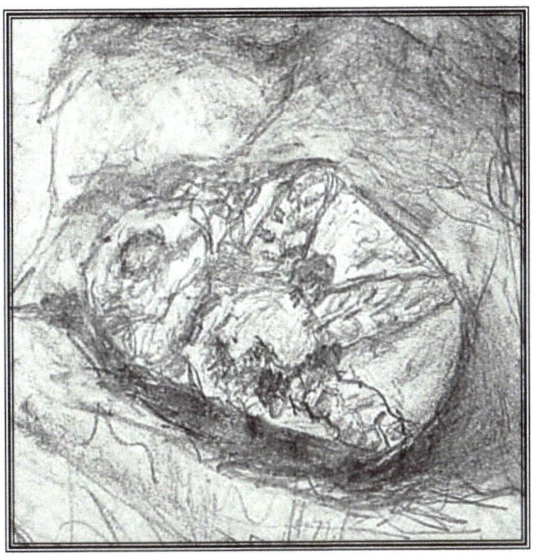

Close up of arm 30x

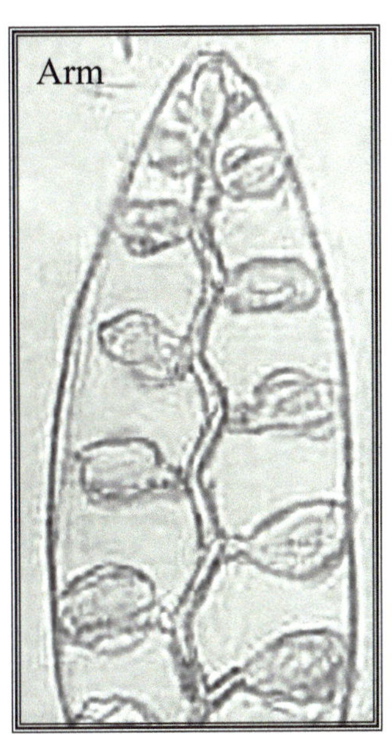

Echinoderms from the Kinzer Formation

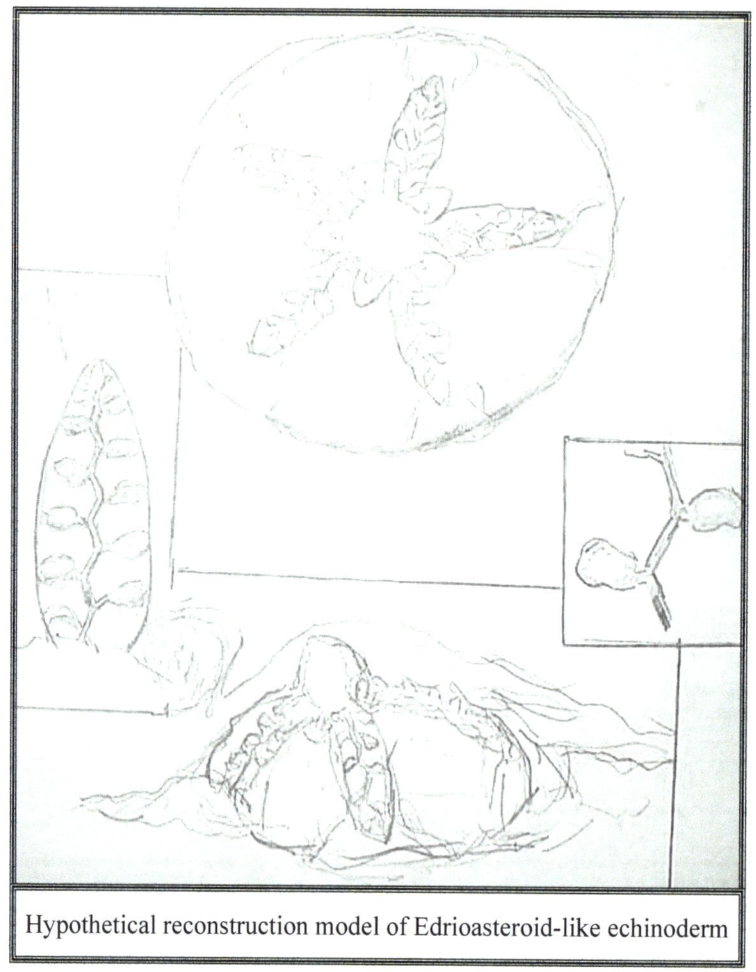

Hypothetical reconstruction model of Edrioasteroid-like echinoderm

A likely candidate for a specimen of Lepidocystis

Echinoderms from the Kinzer Formation

I found little in the way of echinoderms in the rich faunae from areas where I collected soft-bodied specimens near Neffsville. However, one rather peculiar specimen did turn up. It did not seem to appear like *Camptostroma,* but rather a Cystoid-type animal. It has a spherical and plated appearance with a few extended extremities protruding from its small globe-like structure. This too could be a possible candidate for *Lepidocystis.* It may be a part of a detailed portion of this animal. The individual plates are larger in size having a detailed signature pattern. It appears as though arm structure may be superimposed on the body surface.

Conoidal Shell Casts found in the Kinzer Shale Formation

My first experiences collecting fossils in the Kinzers began with field collecting visits to the shale pit known as the Fruitville Quarry (now a college lab site). Collecting there produced a fair abundance of trilobites. Also abundant were tiny cricoconarida-like molds appearing as tiny bullets known as salterella acervulosa. This was in the upper member of the Kinzer formation. Salterella was also extremely abundant in colonies at the Long Farm construction site, primarily in sandy layers of the lower member. At the Long Farm Construction site salterella seemed sparse among the prolific trilobite beds in the upper member. The Harrisburg Pike, Landisville, Getz woods site produced very little if any conodial molds of this creature. Salterella apparently lived in specific environments difficult to put into reasoning. For example, why were they abundant in the upper member (in beds having abundant juvenile trilobites) at Fruitville but barely present in identical type beds at the Long Farm site only a few miles northeast? Other cone shaped fossils occur in the Kinzers as well as salterella. One type known as Hyolythies has been found and described from the Fruitville Quarry and other sites. I found one example at Long Farm Construction site and a few finer examples from the Neffsville site. This was usually three to four times larger than the conodial salterella. It had a bisymmetrical pattern in its structure sometimes occurring with an operculum hinged near the aperture similar to conularians. I have seen documentation of well-preserved soft-body parts of Hyolithies. A few of these portrayed a more detailed taxonomy showing a pair of antenna like Extremities protruding from the aperture. Most peculiar to me, and considerably rare were ribbed cornucopia shaped, conical molds of an entirely different species. These minute specimens were found in the fine grain rocks of the Neffsville site. All of these animals were presumed to be similar in nature. In theory, these shells once had soft-bodied organisms living within. The affinity of these creatures may not be precisely known. One could assume the pointed ends of the carapace could be attached or submerged in mud. Soft feeding parts could have protruded through the aperture as theorized with conulariids. Perhaps they also could have been more like coelenterates of the class scyphozoa. Other hypotheses could portray salterella and like species belonging to the phylum mollusca as being gastropod-like.

Salterella 30x

Salterella 10x Neffsville area

Conoidal shaped shell casts, Kinzer Shale Formation, Neffsville

Lower Cambrian Kinzers; Brachiopods, and a rare Plecyopod

As an avid fossil collector in Pennsylvania I soon came to realize the most abundant fossils were shell remains of brachiopods and pelecypods. Fossil rich beds of the Devonian Mahantango, and Ordovician Martinsburg Formations yielded so many specimens I normally tossed most of them away. Collecting in the Kinzer formation was quite different. Being much older it soon became evident marine shells were rather sparse if seen at all. An early reference I used listed only one species of brachiopod from the Fruitville Quarry site called *Obolella.* I have never seen or collected any samples there.

I heard reports of abundant brachiopod specimens occurring in York County. Reports were of specimens primarily collected from the coarse sandy beds above the Ledger Dolomite in the lower member of the Kinzers Formation. It was not until collecting at the Harrisburg Pike Locality, where in the same zone (a sandy bed referred to as the *bonnia* layer) I found an occasional tiny scalloped (angel wing-like) brachiopod. For all general purposes a brachiopod was a prize find to add to the collection coming from the Lower Cambrian faunal.

In later years as I progressed into the Neffsville site, brachiopods appeared in excellent detail yielding a variety of species I wouldn't have imagined coming from the Kinzer Formation. There was one fairly abundant species, *Nisusia festinata,* reaching nearly an inch in size. These appeared much like commonly familiar types from later periods. *Nisusia* had uniform shape with scalloped surface grooves of consistent proportion. Other unusual varieties appeared having more complex (fancier) looking shapes. One species as has a pointed peak near the center hinge. The most common were small usually dark carbon oval shaped specimens with well-defined growth rings.

Most unusual was a specimen certainly appearing to be a Plecyopod. The specimen seems to be well preserved. It is definitive enough to assess it to be of this class. The pelecypod may be a species known as *Helcionella.* It appears fossils of this group only occur in certain environments. Bi-valves of this size most likely lived in an organic rich, calm, shallow water within soft muds. The Neffsville site may exhibit such an environment for the lower member of the Kinzers with its rich faunal including larval trilobites and preserved soft-bodied animals. Other sites where brachiopods occurred in sandy beds may indicate swifter current deposition such as a shoreline environment.

Brachiopods Harrisburg Pike Site

Brachiopod, Paterina bella? Neffsville Site 30x

Lower Cambrian Kinzers Brachiopods, and a rare Plecyopod

Nisusia festinata Neffsville 5x

Paterina? 10x Neffsville site

Brachiopod with raised hinge 30x

Pelecypod Helcionella? 10x

Annelida from the Kinzers and a new taxonomy for Pelagiella

The Neffsville site continued to unfurl its mysteries. Soft-bodied forms and worm-like creatures were not to be ruled out of the possibilities. There were a number of suspect specimens to consider. Some of the examples collected may not be completely ruled out as gut remains or coprolites from other kinds of animals existing in the Lower Cambrian. A few examples seem to have an articulated organic appearance. One specimen was of great exception. On one field trip I discovered a piece of the fine-grained light colored shale with numerous, small, peculiar looking dots distributed over the surface. It appeared this one was worth inspecting through the microscope. To my delight, the small dots (about the size of this letter o) did indeed have repetitive organic structure. Each had the form of a distinct spiral. It was later confirmed this was a species known as *Pelagiella.* (Pronounced pa-lay-gee-ell-a).

The next sequence of events was to cause much attention to this group of specimens. While observing with vidio-photomicrograpic equipment on a screen monitor I began to notice peculiar wispy brush-like tentacles protruding from the aperture of many specimens. Upon consultation with local paleontologists it appears this was a new (never before recorded) taxonomy for the species. Now the big question is what class to put the specimens into. As I write research is in progress though I have a few hypothetical opinions of my own.

It appears some specimens exhibit the brush-like extremities as paired. I was thinking about a type of arthropod, ostrocods being an extinct cousin to modern barnacles. Barnacles and ostrocods are both bi-symmetrical creatures encased in a carapace or shell. Both have paired feeding arms extending from the aperture. Could this be an arthropod living in a spiral-like carapace? Then there are cephalopods. One of the photos of a specimen appears almost like a nautilus with tiny arms extended. However I am thinking this is not likely in the Lower Cambrian. What about a gastropod with some kind of a feather like antenna? All of these are problems based on observation attuned to hypothetical affinities. My best educated guess goes back to logic. I am thinking polychaete worms. The tiny spirals could represent a carapace or do they? Many types of marine worms do create spiral secretions as they grow. Many types indeed have pairs of feather like feeding apparatus. These specimens could even be trace fossils where sudden burial could cause the living worm to retract spiraling into its hole on the sea floor leaving brush marks. Another theory was these animals are actual tubeworm fossils. In this case they were buried in their life-like state. Could there have been volcanic activity in the region producing poison gases or toxic ash clouds suffocating and preserving these along with many other delicate creatures in a fine-grained deposit?

Millimeter size spirals on Kinzer shale

New taxonomy for Pelagiella (photos at 30xand 10x)

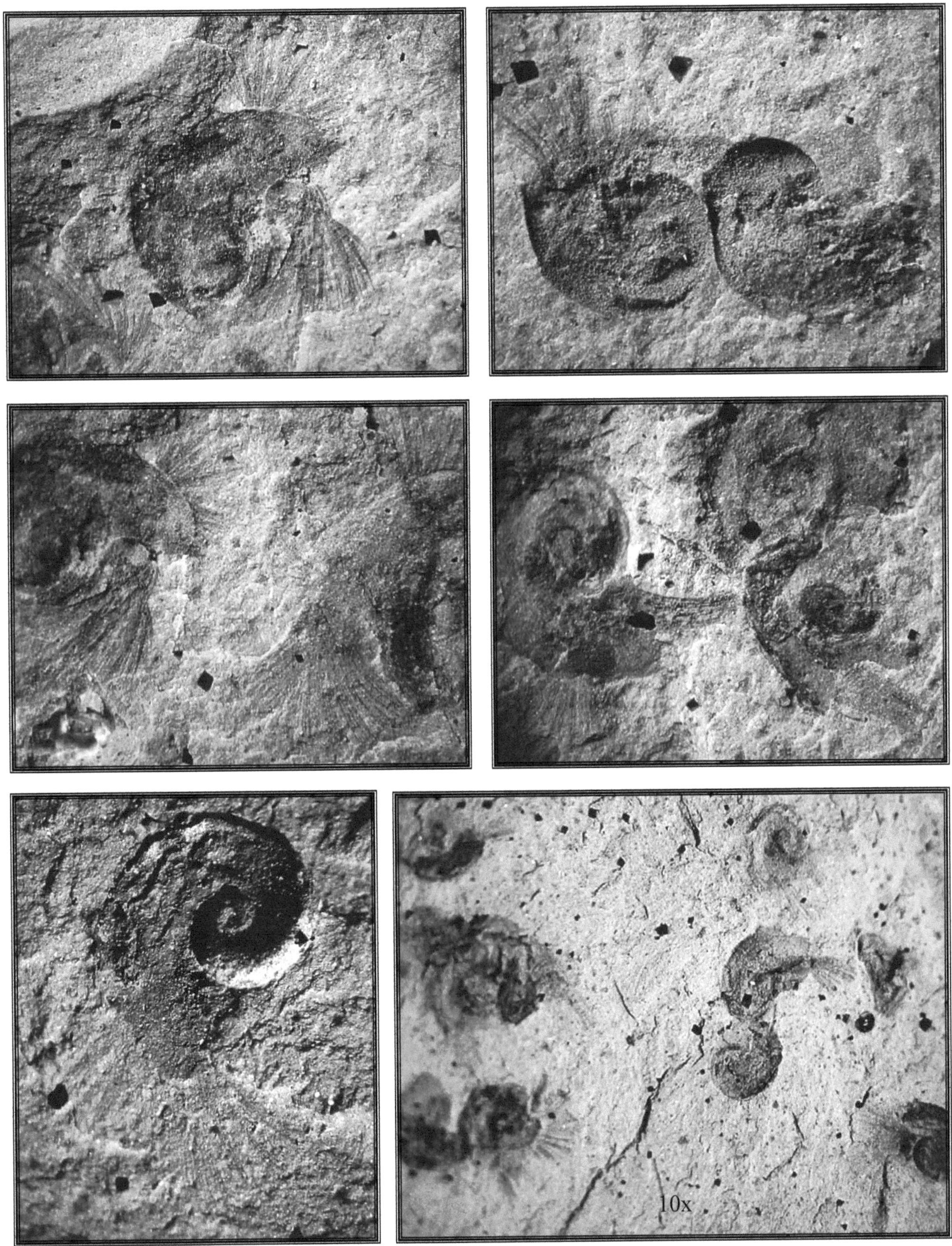

New taxonomy for Pelagiella (photos at 30xand 10x)

Pelagiella exhibiting detailed structure 30x

Example of tiny polychaete on crab leg

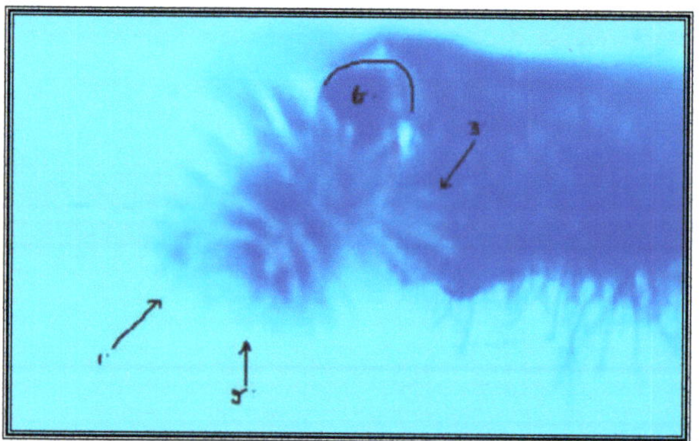

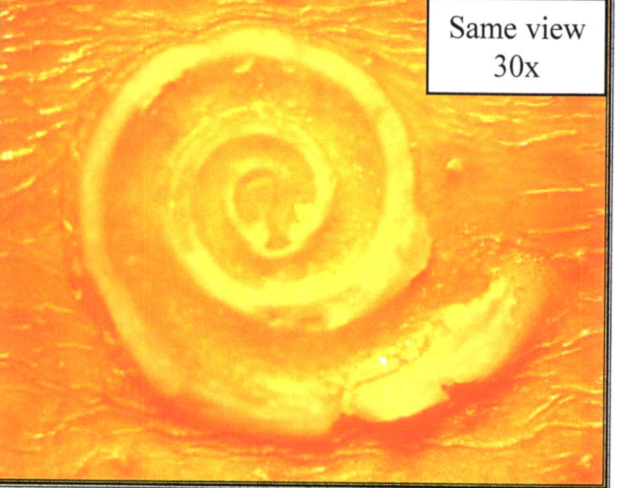
Same view 30x

Modern marine worms showing paired feeding extremities

Pelagiella 30x extremities folded

Potential candidates for annelid specimens from the Kinzers

Lower Cambrian Porifera found in Kinzer rocks

If there weren't enough strange forms from the shale's of the Neffsville site to contend with, there emerged another bizarre creature whose structure was preserved in shale.

My initial thoughts were this being another kind of strange shellfish. I was putting the specimen together with the brachiopods. I had overlooked its imperfection as not being entirely bi-symmetrical. I called it the "corn-digger" since it reminded me of a food snack, having an elongated, somewhat trapezoid shape (perfect for dipping). The specimens I found had a sort of grated, or better yet a tight woven fabric-like appearance on the surface. It was later suggested to me to be a likely candidate for a sponge.

I observed a few specimens of good sponges found in dark slatey shale. They came from a different local Kinzers site having rounded shapes with well-defined spicules. They were fine examples of a different species found by another local enthusiast.

10x

10x

Lower Cambrian Kinzers Coelenterates or Plants?

In addition to the many interesting specimens collected from the fine-grain light shale's at the Neffsville site were curious, repetitive organic patterns resembling marine algae forms. It was suggested from others making observations, various specimens could possibly be coelenterates. Coelenterates are groups including modern marine animals such as sea pens, sea fans, corals, and branching polyp groups. One specimen has a detailed ribbon-like shape. It resembles seaweed exhibiting detailed cellar structure under magnification. Others appeared to be multi branching algae forms shown as fine carbon filaments on bedding planes of shale surfaces. All were true fossils and not mineralization such as dendrites, sometimes confused as plant fossils by novice collectors. Most peculiar were tiny spider like whorl patterns which at first I thought might be iron residue. These patterns seemed to be repetitive. They appear to be of organic origin, some showing better detail than others. There are specimens exhibiting a leaf like whorl, with a stalk or emerging branch, raised from the center, several times the length of the base structure. Could this be a form of marine plant life or an actual coelenterate much like a sea pen?

Branching Algae-like forms

Detailed leaf like blade10x

Branching plant-like structures, or coelenterate?

30x

Lower Cambrian Kinzers Coelenterates or Plants?

10x to 30x

The grand picture in conclusion

I have reviewed my fossil collection of lower Cambrian age animals from different sites within a few miles of my home. For me, collecting such specimens represents something beyond a mere hobby. I like to approach this from a scientific perspective; however ideas I have proposed are not declared as conclusive scientific data. The account profiled (based on many years of acquired knowledge through consulting, internet, museum collections) is meant to be a personal travel log in my journey through the Kinzers. The back yard idea from my title demonstrates the asset of living so close to such prolific fossil environments. My personal opinions, theories, and observations do not intentionally imply conclusive evidence to the precise nature of the specimens exhibited and are by no means intended to infringe upon academically established documentation. Hopefully this will provide supplementary data to be shared with the scientific community. Essentially, two or more heads are always better than one. Statistics gathered from data on many observations including possible new species help add to the grand picture of what our world was like in its early stages of evolution. This also helps us to place where humans stand in time, in this picture. I hope I have been able to share something of pre-historic treasure with this publication

To summarize all the data given creates something for the imagination. Profiling to some degree what the planet might have been like during the lower Cambrian age, particularly here on the local front. If I were to express my imagination I envision the Lancaster area being submerged in a shallow marine environment. At night the stars look quite different, why? The continent was then in a different position on the globe, probably in a tropical latitude. In the distance I envision a range of high snow-capped mountain peaks through the primeval haze across the water. One thing noticeably missing, there are no trees or plants of any kind on a barren moonscape land (if observing along a shoreline). I imagine near-shore waters could be similar to a bay area, (like Chincoteague Va.), where coves of shallow water exist near mainland environments. Perhaps a deeper water shelf off the sandy coast where waves crashed from heavy tidal forces caused by a moon appearing larger, being closer to earth than it is now. This is not the Atlantic Ocean we are familiar with but rather the Iapetus Ocean. I imagine this from observations made at various sites based on types of marine life that existed, also deposition of various sediments. The depth of the sea bottom changed over time during the Cambrian from sediments rushing in, then depositing on the ocean floor. Variations in the strata could represent seasonal rushing of inland melt-wasters from ice and snow on the peaks of a high mountain range. Strata of interbedded fine-grained and coarse-grained sediments could represent seasonal deposition. The deep shelf areas of the Kinzers well profiled in York County show subtle limey deposition where limy sediments of organic origin accumulated and deposited gently in deep water. Many shaley areas found distributed throughout Lancaster demonstrate turbidity where various degrees of sediment filtered into shallower water environments. I picture this (based on specific observations) as similar to the oceanfront, with shoreline and bay. For example fossils from Fruitville quarry, where the upper Kinzer member suggests a quiet environment. Well sunlit shallow waters similar to a back-bay environment could have presented ideal conditions allowing abundant larval trilobites to exist. It seemed many articulated and abundant specimens existed in such environments filled with algae like plant life. A well-nourished sunlit environment might have been well suited for feeding and protection against larger predators. This section could have been a hatchery where larva once flourished. I observed a sample having several minute juvenile specimens of *Olenellus* following each other in a chain as if to suggest trailing through a channel of algae on the muddy floor

The long Farm construction site also appears to show the same stratigraphic profile where there are many juvenile and multiple trilobite specimens in the upper member of the Kinzers like at Fruitville. The LFC site however seemed to produce larger numbers of juvenile *Wanneria* than Fruitville accounting for statistical species difference. The same holds true though with modern marine life forms in environments having serendipitous distribution. Species statistics per-say, may not always relate directly to stratigraphy.

The grand picture in conclusion

The lower member of LFS seems to represent deeper water where turbulenance deposited coarser grained sediments over time. The bedding profile was rather thick showing intermittent clastic variables. (There were sandy layers, and in between, hard, bluish, fine-grained sediment at times bearing pyritized trilobite remains). This pattern was repetitive through most of the depth at this site. Within those beds it seemed there were abundant remains of larger trilobite parts. *Wanneria* seemed to be more common than *Olenellus Thompsoni.* In the rocks due southwest, the lower member from the Neffsville site was filled with fine clastic sediment suitable for preserving soft bodied animals. It was even suggested sediments from the light colored shale there could have been of pyro-clastic origin. If so, this could have poisoned marine life prior to burying the remains. Perhaps there could be much confusion in determining water depth of some marine environments within the formation based on fossils alone. An environment like that of the Neffsville site found in pristine condition could have easily been corrupted by turbidly deposition but apparently was not. Heavy sediment mixed with disarticulated large portions of fossil material might have easily caused lack of preservation of articulated soft body animals, which probably happened within the lower member at the Long Farm Site. The lower member at LFS exhibits an extreme abundance of salterella. It is situated closer to the Vintage Dolomite were disarticulated trilobite parts from *Bonnia* come from a seemingly turbid environment. Throughout the entire formation at LFC site were *Camptostroma*, which were very abundant in the upper member. The Getz Woods site (further west) seemed to represent a much different environment. I propose the idea of deeper water where finer material could suffocate living organisms producing lithography of a different clastic nature. This shale was a gray hard porcelain-like material yielding excellent fossil preservation. Many specimens collected from the site were well-articulated rather large examples of mainly *Olenellus Thompsoni.* Juvenile trilobites seemed less common but still had reason to be carried in by currents. There seemed to be a lack of any types of mollusca at Getz Woods Site. The Harrisburg pike site displayed a different lithography a few thousand feet northward from Getz. The site also produced large well-articulated trilobites. Here the bedding was different from most known Kinzers shale's. The environment seemed well suited for the trilobite *Wanaria Walcottana* where many large articulated specimens were collected. A close stratigraphic profile seemed to prevail within the lower member.

The bottom adjacent to the Vintage Dolomite was a coarse sandy material with poorly preserved brachiopods and abundant *Bonnia.* The layer above yielded many specimens of *Camptostroma*, *Olenellus Thompson*i, and *Mesonachis-like* types in a silt-like rock bearing reddish oxide. The upper layer was a light yellowish punky-like bedding of fine sandy sediment, producing many fine *Wanneria* specimens. Each layer was less than a foot in thickness. Throughout the Kinzer formation many different conditions can be recorded as we have discussed. Another item I need to mention is slump deposits. We can't rule out the possibilities of quick burials of marine life from a collapsing ridge weather it be along a trough or a cliff base. Determining an exact map of how our local Cambrian appeared is somewhat difficult since the sea floor was constantly changing over millions of years. It is a struggle of stratigraphy versus the paleo-biology yet something of a picture can be painted through logic based on statistics. While visiting mud flats in the Back Bay at Chincoteague I often made observations comparing them to paleo-environments. In conclusion I look at what I see in the Kinzer environments similar to those I see in the modern scenery.

A deeper ocean with a sandy shoreline, shallower back-waters with turbulent tidal flows carrying and constantly depositing sediment from heavy sands to soft mud. Rich muddy basins harboring marine vegetation, delicate organisms and tiny mollusks. I envision a sandy shoreline during the Cambrian period where there were abundant *Skolithos* (borrowing tube worms). Large trilobites were washed on the beach rather than horseshoe crabs. The only sound is crashing waves on the shoreline with endless moonscape like horizons. The atmosphere is extremely low in oxygen. Nevertheless the surrounding marine environments in the calm backwaters were exploding with abundant strange life forms. "our oldest ancestors".

Lower Cambrian Kinzer Shale sites

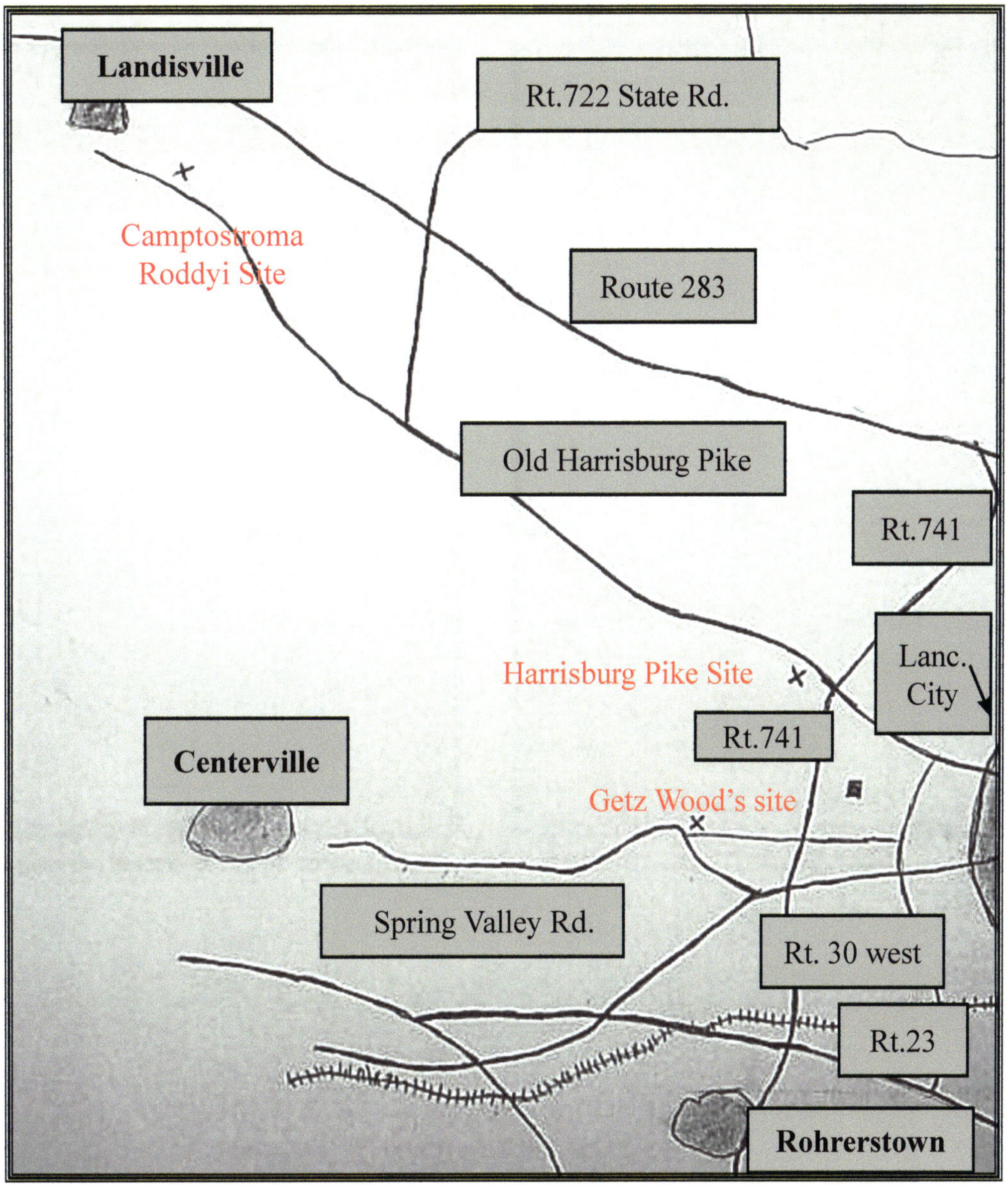

Landisville

Rt.722 State Rd.

Camptostroma Roddyi Site

Route 283

Old Harrisburg Pike

Rt.741

Lanc. City

Harrisburg Pike Site

Rt.741

Centerville

Getz Wood's site

Spring Valley Rd.

Rt. 30 west

Rt.23

Rohrerstown

Pit in Getz Woods

Harrisburg Pike Lower Cambrian Fossil Site
(Photo Around 1986)

Excavation Pit (Kinzer Shale)

Fossil site distribution: Kinzer Formation, Lancaster Co.Pa.

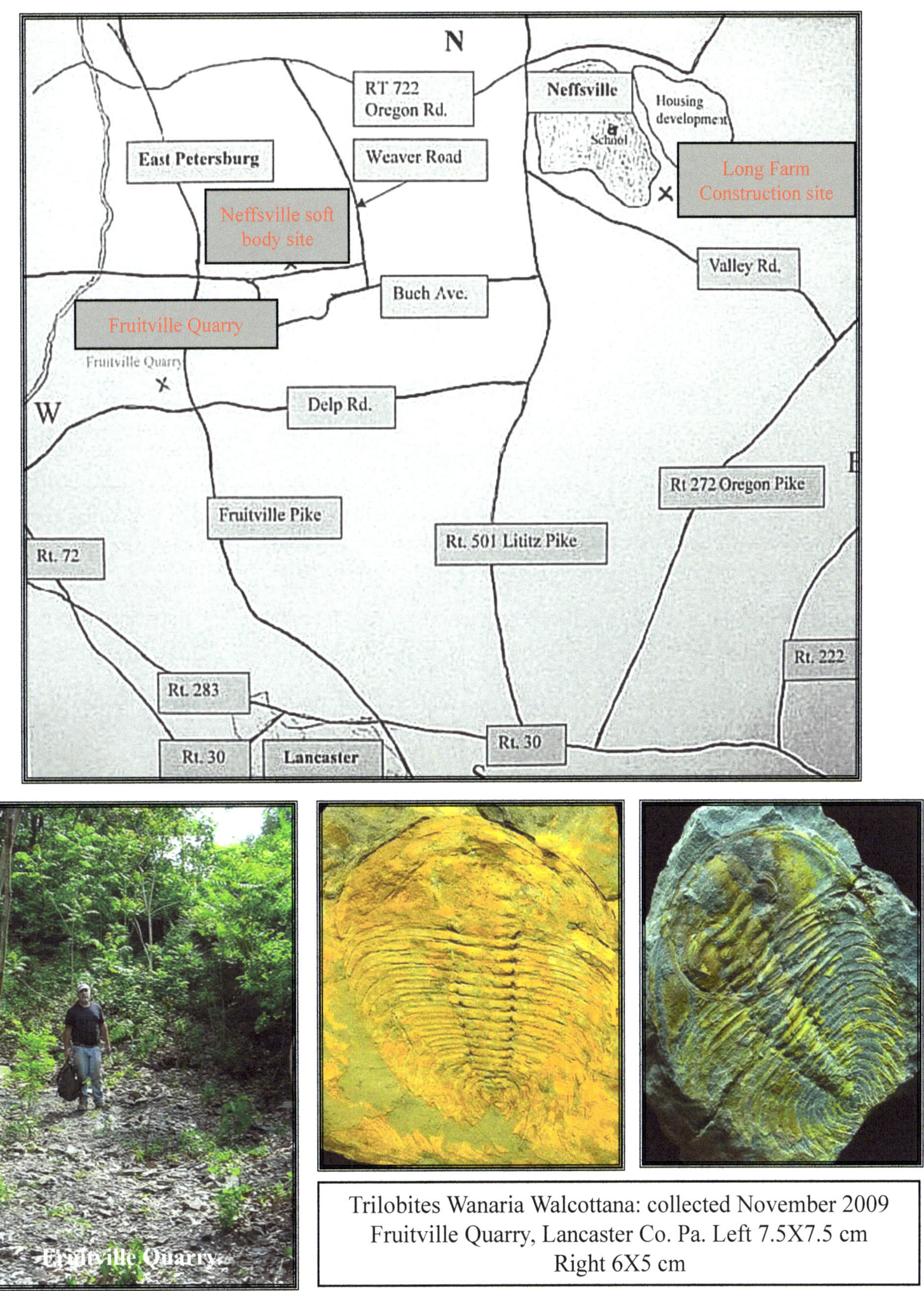

Trilobites Wanaria Walcottana: collected November 2009
Fruitville Quarry, Lancaster Co. Pa. Left 7.5X7.5 cm
Right 6X5 cm

Lower Member bedding at Long Farm Construction Site

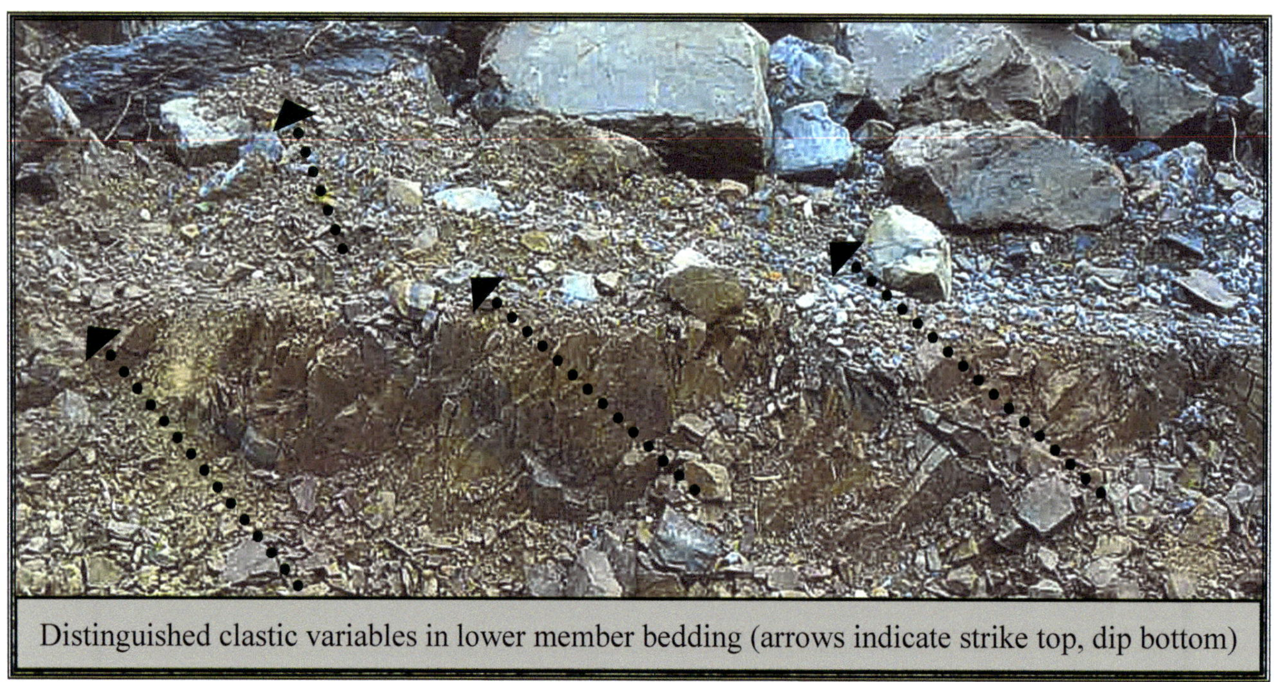

Distinguished clastic variables in lower member bedding (arrows indicate strike top, dip bottom)

Sedimentary clastic variables, hard finegrained blue rock, notice the pronounced band of a porous sandy layer in lower boulder is retaining moisture

Upper Member of Kinzer Shale Beds at Long Farm Construction Site

Rich trilobite beds in upper member

Freshly exposed 5" Wanaria Walcottana, first light in 520 million years Rich trilobite beds in upper member

Acknowledgements

Paleontologists ()=original discoverers of species appearing in text

TRILOBITES
Ollenellus thompsoni (Hall)
Ollenellus vermontanus (Hall)
Ollenellus fremonti [western US] (Walcott)
Ollenellus Clarki [western US] (Reeser)
Ollenellus Getzi (Dunbar)
Paedumias Transitans (Walcott)
Mesonacis (Walcott)
Wanaria walcottana (Wanner)
Olenoides [species of specimen in this book un-verified]
Kootenia {species prici} (Reeser)
Bonnia {specimens in book may be *Bonnia occipitalis*} (Rasetti)
Lancastria Roddyi (Walcott)

NON-TRILOBITE ARTHROPODS
Anomalocaris { Pennsylvania species} (Reeser)
Canadaspis [Burgess shale British Columbia Canada]
Sidneyia (Walcott)
Serracaris Lineata (Reeser & Howell)
Wiwaxia [Burgess shale British Columbia Canada]

ECHINODERMS
Camptostroma roddyi (Ruedemann)
Lepidocystis {species wanneri} (Foerste)

MOLLUSKA
Paterina [species of specimen in this book un-verified]
Helcionella {species prici}(Reeser & Howell)
Nisusia festinata (Billings)
Obolella [from field guide bulletin Pa.G-40 1969]

OTHER
Pelagiella {Species exigua}(Reeser & Howell)
Hyolithies {Species americana or communis} (Billings)
Salterella {Salterella acervulosa} (Reeser & Howell)
Hallucigenia [Burgess shale British Columbia Canada]
Odaraia [Burgess shale British Columbia Canada]

{} = Possible name of species appearing in this book but not verified.

Directory, Lower Cambrian Explosion In My Back Yard

Introduction	1
Trilobites Olenellus Group	2-9
Mesonachis-Like Trilobites	10-11
Trilobite Wanaria Walcottana	13-18
Trilobite Bonnia	19-20
Rare Trilobites	21
Rare Arthropod	22-24
Phyllocarid	25-26
Serracaris Lineata	27-28
Anomalocaris and other rare arthropods	29-31
Echinoderms	32-39
Conical Shell Life	42
Brachiopods and Pelecypod	43-44
Pelagiella (New Taxonomy)	45-47
Annelids	48
Sponges	49
Coelenterates or Plants ?	50-51
Analogies	52-53
Site Photos and Descriptions	54-58

All specimens portrayed in this publication were self-collected by Kerry Matt at locations described, from 1970's up to the time of this writing. All Photography, drawings, and editing: by Kerry Matt. Special Thanks! To Jim and Mimi Stauffer for helping make the long farm construction site accessible. Special thanks: to all landowners allowing collecting on their properties, and the companionship of fellow mentors / collectors.

Lower Cambrian Trilobite
Olenellus Thompsoni, Neffsville Pa.

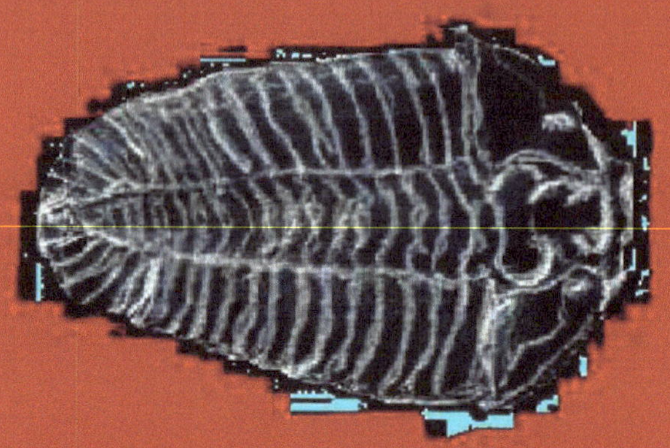

Swatara Gap Pa.

Ordovician fossil fauna

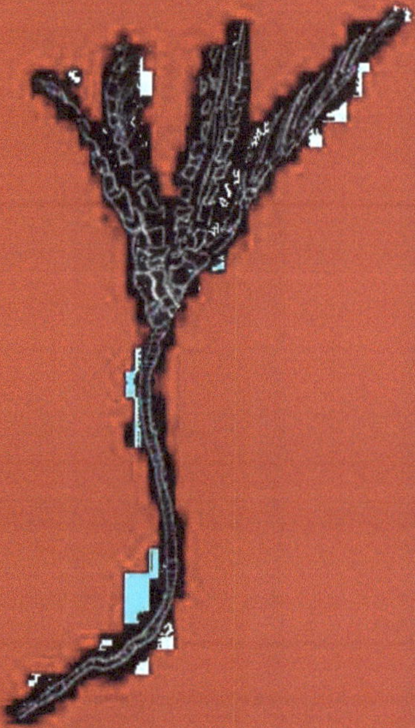

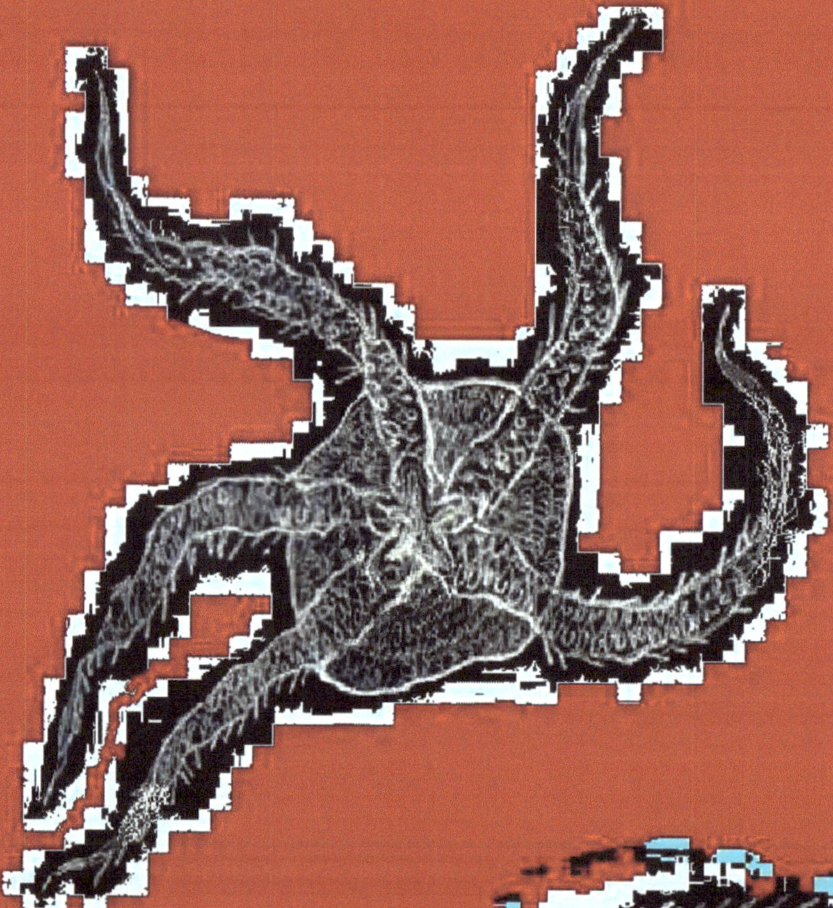

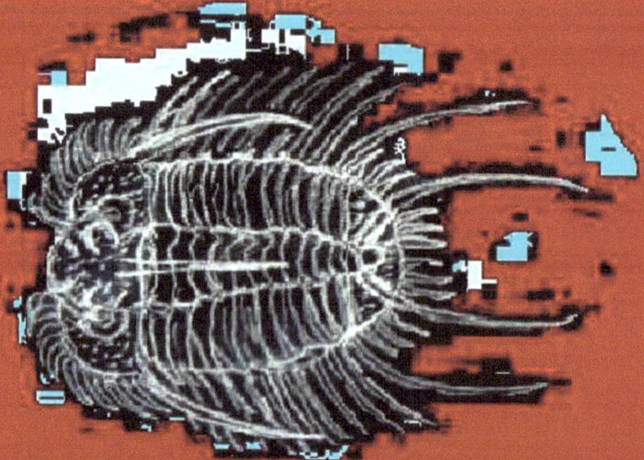

By: Kerry Matt

Swatara Gap site as it appears today

Swatara Gap 440 million years ago

Directory: Swatara Gap Ordovician Fossil Fauna Lebanon Co. Pa.

Index	Page 1-2
Introduction	3-4
Trilobites (arthropods)	**5-22**
Cryptolithus	5-8
Flexicalymene	9-11
Isotelus	12-15
Triarthrus	16
Acidaspis	17
Diacanthaspis	18
Pterygometapus	19
Mesotaphraspis	20
Platylichas	21
Proetus	22
Echinoderms	**23-36**
Starfish	23-32
Taeniaster	24
Protasterina	25
Mesopaleaster	26
Phragmactis	27
Petraster	28
Salterella	29
Schuchertia	30
Promopaleaster	31
Crinoids	**33-36**
Ectinocrinus	33
Glyptocrinus	34
Heterocrinus	35
Echinosphaerites	36
Brachiopods	**37-39**
Rafinesquina	37
Leptaena	38
Dalmanella	38
Sowerbyella	38
Sericoidea	38
Hebertella	39
Lingula	39
Pholidops	39
Craniops	39
Plectorthis	39
Pelecypods (clams)	**40-41**
Gastropods (snails)	**42**
Other common species	**43-44**
Annelids (worms)	**44**
Graptolites	**44**
Cephalopods	**44**

All drawings by Kerry Matt are scaled to approximate average adult size of each sketched individual

Directory: Swatara Gap Ordovician Fossil Fauna Lebanon Co. Pa

Rare Specimens	45-46
carpoids	45
Conularians	45
Lichenids (speculative)	46
Eurypterids (arthropods)	46
Conoidal shell	46

Paleontology is an ongoing cutting edge science: This documentary is a personal field log not intending to declare absolute or final scientific data. Analogies presented in this publication are hypothetical based on my personal philosophies. Names of all specimens are ones used in the field and lab for identification. In some cases a species name might have been updated. Some names in this text might not be accurate to present date.

Pseudo star (Crinoid)

Pseudo star (Crinoid)

Uncommon brachiopod

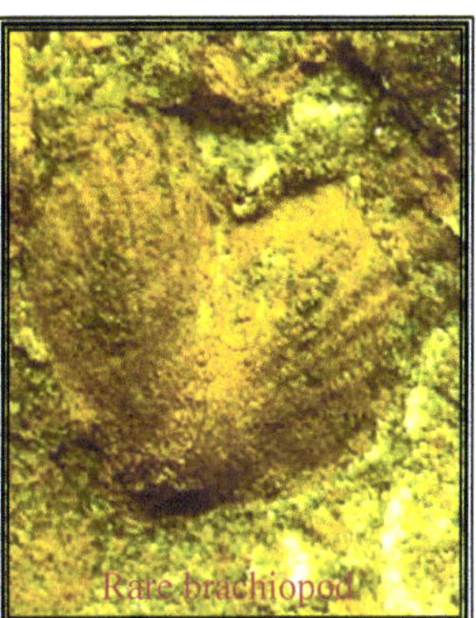
Rare brachiopod

This book is dedicated to all friends in company whom enjoyed this site with great enthusiasm while it lasted. Definitely, A World Class Fossil Locality! Kerry Matt

Starfish Phragmactis

Swatara Gap, Lebanon Co. Pa

World Class Ordovician Fossil Fauna

For me fossil collecting started in the 1970's as a member of our local colleges Jr. Earth Science Club sponsored by the North Museum of Lancaster Pa. I reminisce and treasure times of past adventures to the famous site with our club. Our mentors presented a chart exhibiting diagrams of various recorded species for on location reference. After our arrival, soon after a briefing, members scattered about on the quarry benches and talus piles to find the nicely preserved steinkerns the site was noted for. Vivid yellow and orange oxide coatings gave the fossils a distinguished signature appearance for this locality. A heightened objective was voiced to find rare and elusive fossil brittle starfish. Brachiopods were immediately perceived as abundant as well as trilobites, mostly a plethora of assorted disarticulated fragments. Some members of the club seemed content breaking rubble while I preferred chiseling away at bedding along a ledge. My find of the day was an inch long segmented trilobite with brown oxide coating. Not as common as the trinucleidae species Cryptolithus the site was most noted for, it was a multi-segmented body of the species Flexicalymene exhibiting exquisite preservation with raised relief. After removing the specimen from the upper corner (left of center of the site) little did I realize at that time I could have extracted the rest of the head from the section of rock where I was collecting to be reattached later making a complete specimen. I was new at this having no foresight to my future of expanded knowledge and preparing skills.

The site was evidently a borrow pit for fill shale used for a railroad bed running north and south between the Swatara Creek and Route 72 adjacent to (east of) the site. From the time of my first visit the cut on the west face of the gap, center of the mountain face stood beneath the southbound lane of Interstate 81 where a large concrete bridge raised on tall pillars crossed directly above the lower quarry, Route 72, and Swatara Creek. I have little account of the exact history of the borrow pit often wondering how many great specimens might have been lost, not thoroughly knowledgeable of finds prior to the construction of Route 81.

The formation was identical above the bridge as below. Stratigraphic bedding layers are situated in a nearly vertical position in the mountain face. In the early days collecting I spent most of the time working the lower quarry. In the 1970's and 80's there was plenty of room to drive into the lower quarry from Rt. 72, and park along one of the concrete bridge abutments. I recall school buses full of eager students pulling in there.

In the glory days of collecting, almost every weekend, one became used to the clanking sound of people chiseling away, sharing the enthusiasm of their finds. In later years the lower quarry was filled in with mounds of waste dump rock and fill from other sources. Parking became constricted, mainly limited to the shoulder of the main road. Eventually the state posted the site off limits as an upper section of the lower quarry was eroding too close to a shoulder of Rt. 81 south. The upper level (right shoulder 81 south) on the mountain face was once popular for collecting after the highway department removed old talus. Many rocky ledges became fully exposed. I spent many seasons collecting to compile the most complete fauna assemblage possible.

The geology setting is in a shaley to sandy bedded rock called the Martinsburg Formation dating to around 440 million years age. The clastic nature of the rock seems to reflect weather conditions affecting sediment deposition in a shallow Ordovician sea. Advancing forms of many species of marine life conducive to a well sunlit environment fossilized to become preserved with iron oxide material further oxidizing into steinkerns bearing orange and yellow residue. The prolific fossil unit consists of interbedded layers mainly situated in the core of the exposed face in the western side of the mountain gap. The strike of the bedding apparently crosses the Swatara Creek at a very slight angle northwest to southeast.

Swatara Gap, Lebanon Co. Pa

World Class Ordovician Fossil Fauna

The exposure should potentially be visible in the east face of the mountain gap across the creek but is apparently covered in a slump of quartzite boulders and conglomerate rubble. Facing west to the exposed bedding was a breadth of more or less the length of a football field. An abrupt contact zone occurs on the north side (right side, as you face west into the cut) where the Martinsburg shale meets the Tuscarora Conglomerate.

The Martinsburg bedding was flatter and blocky nearer the contact for several yards south from the contact and rather barren of fossils, however conducive to an occasional rare specimen. The southern side of the exposure merged beneath the bridge at the lower quarry. Some exposed areas farthest south in the formation produced friable shale's yielding fossil graptolites.

The paleo environment suggests buildup of thick muddy sea floors (bluish green thicker beds) separated by blocky, (brick-like tan colored sandy) layers of a different clastic nature. The stratigraphy of the site suggests bedding layers could have remained stable for a duration allowing delicate marine life forms to accumulate later becoming corrupted by sandy sediment carried by turbulent currents, possibly an effect of seasonal weather conditions. It is also suggestive slumping may have occurred with evidence of rolling of clay lumps (concretions) on the sea floor. The sea was most likely gentle, shallow, and well sunlit in order to provide a well-nourished environment conducive to many delicate species described. I will attempt to describe all of the most interesting specimens placing each one in order from most common to most rare including the paleo-environment of each type according to field statistics.

(4)

The Trilobites of Swatara Gap

Trilobite, Cryptolithus Bellulus:

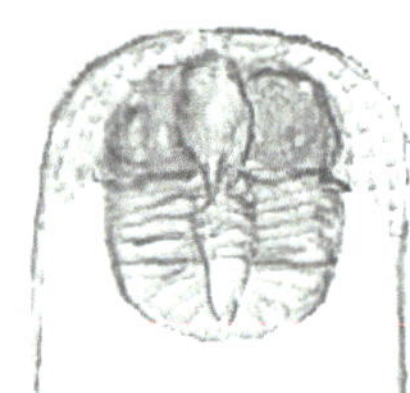

Making the site popular are exceptional specimens of a trinucleid variety Cryptolithus bellulus. These trilobites are unique in appearance compared to the general taxonomy of typical varieties. They are oval in shape having a large bulbous glabella (pseudo nose plate) with smooth surface. On either side of the glabella are large circular patches as eye lobes with continuing flat smooth surface (usually shiny and bearing no orange oxide). I have theorized, (though many paleontologists conclude these animals were blind), that these could have been clear shield plates like large headlight covers. The soft eye tissue might have been placed beneath as if looking through a glass shield. The cephalon bordering the head is a pitted collar often bearing long thin genal spines exceeding the length of the entire body. Perhaps these spines could be used as anchors to secure the animal in the muddy bottom during times of swift current activity. The circular six-segmented thorax is bisected with both subtle narrow axial spine and plurons tapering down at the ends creating a fine edge. A small narrow pygidium (tail) as a semi-circle rounds out the entire body bearing subtle fan like furrow ribbing. Many specimens are often found with no trace of genal spines suggesting a possible difference in species or gender.

Cryptolithus is extremely abundant throughout the formation from minute larval stage (juveniles) to adults reaching about an inch (maximum size) body length. Many disarticulated fragments and sections are commonly found throughout most of the fossiliferous unit. Head shields are popularly collected and easily spotted as they retain a great amount of orange oxide on the pitted surface. It may be suggested, these were free-swimming arthropods living in all parts of the paleo-environment. Some of the thick bedded sandy layers, particularly in the southern most quarter of the formation, were filled with many head shields (hash beds—many on a small block) and at times articulated groups occurred.

{I can attempt to describe where certain bedding layers were located by bisecting the unit with imaginary boundaries. Lets suggest a line bisecting the middle of the site as we face west into the mountain with our backs toward the Swatara Creek. Next we will divide this into the southern two quarters (left) and the northern two quarters (right) we can use this reference as we work through the faunal description of each species.}.

In bedding of a pale brownish deposit from the northern most part of the first (farthest south) quarter, complete specimens of Crypro's (as we called them) seemed abundant. There were times five or more complete specimens could be collected on a days dig. (Typically for me a dig would be 5 to 7 hours at a time) In this layer were abundant brachiopods sericodia and the trilobite Triarthus (mostly disarticulated) all in association with relative consistency occurring in continuous bedding in both the upper and lower sections of the mountain face.

Some beds yielded better preservation based on clastic composition. It was easy to accumulate a large number of these while in pursuit of the more rare specimens over time. Specimens of Cryptolithus were also common in the darker slatey bedding in the center of the quarry face. Bedding around the southern edge of the northern most quarter produced large excellent quality specimens in thick greenish blocky shale's with other more common trilobites of the Ordovician. In general Cryptolithus could be found having good quality and articulation in most any layer throughout the entire unit.

Trilobite: Cryptolithus, Swatara Gap

(6)

Trilobite: Cryptolithus, Swatara Gap

Trilobite: Cryptolithus, Swatara Gap

(8)

The Trilobites of Swatara Gap

Trilobite, Flexicalymene Granulosa:

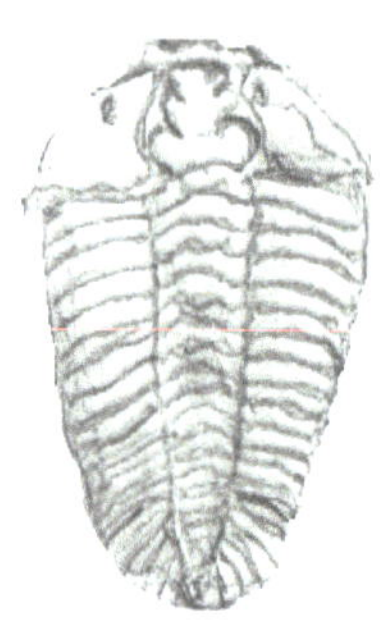

The second of the more common (and highly praised for this locality) has a most typical appearance. This is a trilobite exemplifying one the average person imagines a trilobite usually appears to be like. It has a distinguished multi-faceted cephalon with neatly sculptured knobby glabella including free cheeks attached to the cephalon. The eyes are small (dot like protrusions) well positioned on the head shield. Flexy's have a longer; larger heavily ribbed thorax and axial spine (looking like a vertebrae column with about 12 segments not counting those within the pygidium). The body is usually broad from the first segment beneath the head gently narrowing toward the tail. The pygidium is barely distinguishable from the rest of the thorax.

They are usually found well preserved as steinkerns with vivid raised relief. It seems these animals were scattered bottom dwellers moving about the sea floor much like a horseshoe crab. This species was abundant in this kind of Ordovician environment, found articulated throughout much of the bedding including turbidity deposits. It is common to find enrolled specimens and molts of this arthropod. Cryptolithus was also often found as enrolled (example: curled up like an armadillo). To find one to three good specimens on an average full day of collecting was about on par. There was a bedding layer approximately in the middle of the 3rd and 4th quarter (north) conducive to finding pods of larger well-preserved specimens of this trilobite along with other typical Ordovician species Cryptolithus and Isotelus. Generally good examples were found mainly in greenish heavy thick beds intermittingly distributed across the litho face of the formation. At times what appeared, as a good complete example would be taken home to be prepared in the lab only to find the head missing. In many cases you would end up having a molt. (All arthropods molt shedding the external skeleton) {One trilobite could indeed make many fossils this way} It seemed these animals shed pulling their bodies through the dorsal side between the thorax and head shield. Often you found bodies and heads close together on the same matrix. Molts could be found in all species of trilobites and mouth plates, (easily confused as a mollusk shell by individuals unaware of this) could also be encountered from various species. A mouth plate was a part of a trilobite (feeding apparatus on ventral side) called the hypostoma. (Not usually visible) Flexicalymene is found from a few millimeters in size up to around an inch and a quarter in adults. This species was a proto-type to larger calymene trilobites later immerging in the Silurian period.

Hypostoma

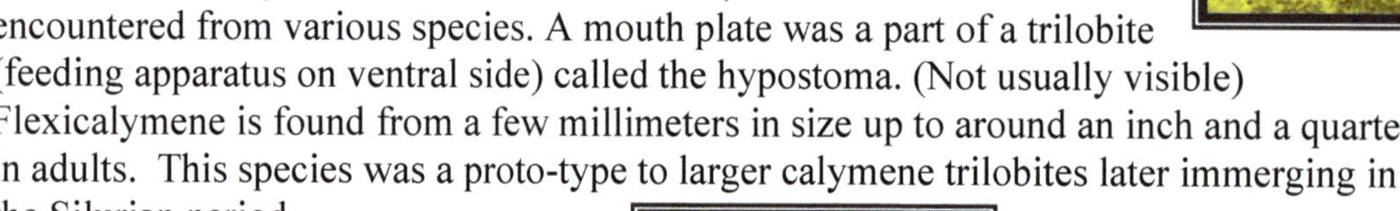

Trilobite molt with exit trail

Enrolled

Enrolled

Trilobite Flexicalymene, Swatara Gap

Trilobite Flexicalymene, Swatara Gap

The Trilobites of Swatara Gap

Trilobite, Isotelus:

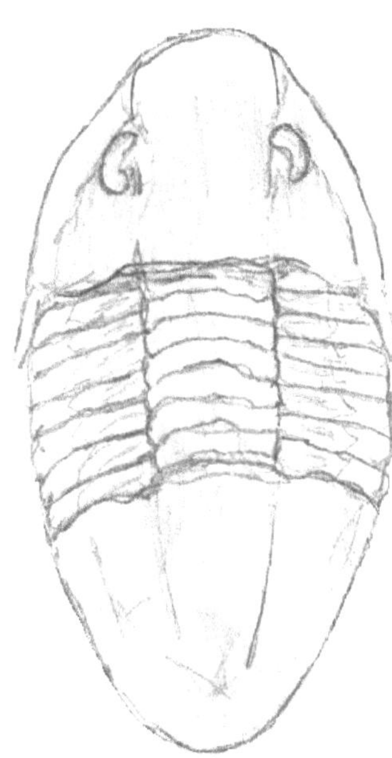

Number three of the most common four major trilobites known to typical Ordovician paleo-environments is the giant in our fauna from Swatara. This trilobite was most prevalent in a thick bed previously mentioned for producing abundant Flexicalymene, and Cryptolithus. As with Cryptolithus, but entirely different, was again nothing like your typical species. They grew to large proportions and could be thought of as an armor-plated trilobite. The taxonomy was certainly different in structure yet indeed, a trilobite. The head exhibits mostly a smooth flat surface. There were lines usually showing attachment scoring dividing the cephalon boarders from the glabella area. All relatively uniform as the likes of turtle shell. The compound eyes are large and raised having a crescent or kidney bean like shape. Proportionally thick genal spines can usually be seen extending from each cheek with a moderate extension length. (Perhaps not extending beyond a quarter of the thorax length) The axial spine is about as equally broad as the lengths of the seven plural sections on either side of the thorax. The thick plural segments taper downward along the length of the thorax to a rounded or paddle-like ending. The pygidium is mainly a featureless smooth armor-like surface of triangular shape resembling a large clam.

I found pygidium of sizes suggesting this trilobite could reach beyond six inches. I believe those less familiar with these disarticulated plates could easily mistake them for large pelecypods.

My best find occurred on April 2nd 1984. I was collecting somewhere near the middle of the first and second quarter high in the formation under a group of trees. The piece of shale frames the trilobite very well and the 7cm long trilobite is for the most part in perfect condition. I prepared the specimen in the lab and it stands up well in its matrix. This was the first of perhaps a half dozen good specimens I collected over the years. There was, as I mentioned, a layer of bedding with the more prevalent Flexy where you could readily encounter disarticulated specimens of Isotelus. I recall a few incidents where potential completes were lost due to the brute forces of the excavating process.

On one occasion I saw the outer parameters of a large specimen near the center of the entire exposure emerging in a protruding cleft of rock. It appeared a bit crumbly.

My collecting skills had advanced and I knew exactly what to do. I carefully excavated everything into a cardboard flat. Every crumb was captured. Sure enough back in the lab after sifting through the dusty jigsaw puzzle I was able to restore the trilobite in its entirety. This was a great salvage effort proving out all five inches of my largest complete specimen of Isotelus. In later years (Before site closing) I again found one in excellent condition just shy of four inches well displayed on a large slab of Martinsburg green shale (near where the one form 1984 was collected). These were generally found in the thicker greenish beds with the Flexy's and Crypto's throughout the span of the litho face.

I envision this animal as having lived its life much like a horseshoe crab crawling or resting on the sea bottom. Its body shape suitable for partial burial in silt or mud with keen compound eyes raised above its cover perhaps in search of prey. Though many trilobites are thought of as filter feeders, sifting through sediments, Isotelus with raised compound eyes along with its beastly size compared to all other local species could be suggestive as a stealthy predator. Another species Hometelus Steggops, described from Swatara is much like Isotelus with minor taxonomy variations appearing one had genal spines, the other did not. I acquired a specimen labeled Hometelus Steggops having a strong resemblance to Bumastus. Based upon past consultation, Bumastus may be present. Bumastus and Hometelus extend trilobites to a dozen different species.

Fossil Trilobite Isotelus, Swatara Gap

Fossil Trilobite Isotelus, Swatara Gap

Bumastus or Hometelus Steggops?

Trilobite: Isotelus, Swatara Gap Lebanon Co. Pa.

The Trilobites of Swatara Gap

Trilobite, Triarthrus Etoni:

This is number four in the count for Swatara trilobites. Through the years the number of disarticulated specimens might put this closer to number two however articulated finds are actually on the rare side. For all general purposes I should almost put it further down the list but Triarthrus is common as a species encountered in other Ordovician sites. As for complete specimens I only have two good ones from all the years collecting.

10 X

A layer in the second quarter seems to commonly produce parts (mainly heads and thorax sections) of this small trilobite found readily with Cryptolithus often with a tiny brachiopod species sericodia. Most specimens are under a quarter of an inch from Swatara. My largest (and also largest I have seen from this site) is two cm. Here again is a trilobite looking like a typical type. The crainidium has broad cheek sections but the actual eye itself is a thin crescent, rather flat, and compound. The head over all resembles the head of a dragonfly. The glabella is usually flattened and grooved on each side. The distinguished signature appearance over all is the short or stubby plurons as compared to the broader axial spine. The long narrowing framework of Triarthrus generally gives it a bug like appearance. Fragments can be found in most any bed to some degree. There were different species recorded abroad, some were listed as Triarthrus Becki and others Triarthrus Etoni. The Etoni species was the one deemed for Swatara's faunal list. On my specimen I count around 17 segments but it is difficult to get a precise count. The general appearance of the thorax seems to merge into the pygidium area with little distinction. This species probably wandered along the sea bottom living a life similar to Flexycalamene.

Approx: 1"

Approx: ½"

The uncommon trilobites of Swatara Gap

Trilobite, Acidaspis Cincinnatiensis:

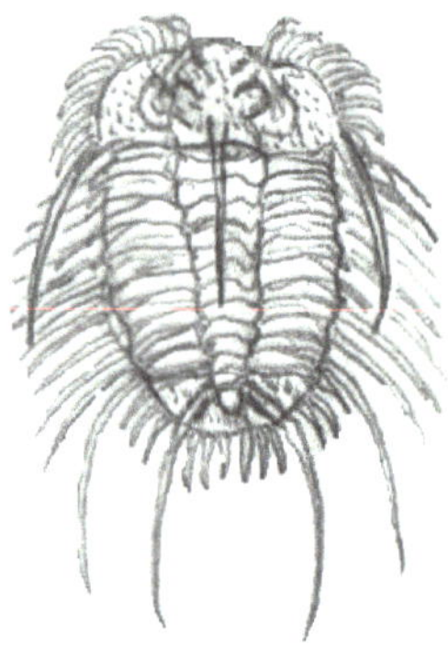

{fifth in place} I will certainly say I was surprised in the early days as a collector of how many unusual species there were to add to the list for Swatara. One species high on everyone's list for the most knowledgeable collector was this unusually complex looking spiny trilobite, and to find one was a prize. To best describe these, a curved spiny boarder (bisected with vacant space just above the glabella) surrounded the crainidium as if to suggest having whiskers, or a main. The interior is pitted in appearance and the glabella area more like an exaggerated form but similar to the glabella of Flexycalamene. The eyes are small, compound, and slightly raised on stalks. The bodies were generally more oval shaped like Cryptolithus. The axial lobe and plurons were somewhere between Flexicalymene and Cryptolithus but proportionally about the same. Long spines extending from the tip of each pluron surround the thorax of Acidaspis. The pygidium similar to Cryptolithus displays a pair of raised bars half the distance from a center point on each side with extra long spines attached. The genal spines extend nearly half the length of a specimen and a third spine (usually missing) protrudes from the glabella. This third spine also shows up on the glabella of Cryptolithus but is generally missing or inconspicuous. I have been lucky but also dug hard enough to encounter a few specimens of Acidaspis. The ones I found seemed to be on the fourth quarter side to the north of the exposure. I have observed a few exquisite specimens form Swatara Gap. The maximum length is about ¾ of an inch or generally under an inch. They could be found in both green shale's and sandy blocky beds most often in disarticulated condition. Good specimens were mostly found in darker shaley bedding. This trilobite was obviously well protected from predators so it probably spent its life exposed. It could possibly attach its self to other organisms, plants, or the sea bottom raising its body having the freedom to feed from the fragile ventral side as water flowed through its surrounding spiny cage.

Actual specimen Approx: 1"

The uncommon trilobites of Swatara Gap

Trilobite, Diacanthaspis:

{sixth in place} Now we are getting into a very uncommon species though quite similar to Acidaspis primarily the difference here is size. These trilobites were often found associated with rare echinoderms and in turbidity hash beds more so than other layers. I only found about two specimens in all of my years collecting. My best specimen was collected from dump piles moved to a park six miles north of the site. Perhaps they were often easily missed because of their small size. Diacanthaspis never got much larger than the eraser on the end of a pencil.

The name Diacanthaspis stems from a taxonomy feature. Pronged spines (usually not intact) protruding back from the glabella are paired (Two = di). Another feature is dotted ornamentation within the plurons of Diacanthaspis as compared to subtle ornamentation on the plurons of Acidaspis. Both species have 10 thorax segments (my best observing skills in tact). Most reports of Diacanthaspis are from the sandy layers in hash indicating they may have been swept in by currents form another paleo-environment of close proximity. They might not have lived abundantly within the local paleo-environment.

Actual specimen is approx: 1cm

The most rare trilobites of Swatara Gap

Pterygometapus (achatella) achates:

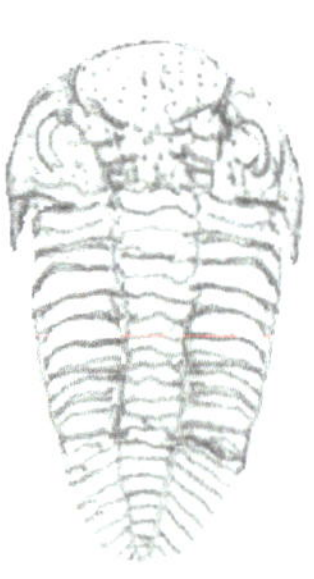

{seventh in place} Pronounced with the P silent as (Terry-go-met-a-pus). This trilobite is elusive to collectors as far as finding complete specimens. To my best knowledge only heads and pygidium were found to any significant degree. A few partial specimens rarely turned up with thorax sections included. Most were found in hash blocks in the pale sandy bedding. It seems as though disarticulated specimens arrived in those beds carried by swift currents from another nearby paleo-environment better suited for living trilobites of this species. It could be similar to a modern bay area where certain crabs are more abundant in certain areas and not found at all in other places, yet fragments of molts can easily be washed out of place by currents. I have only ever found either heads or pygidium of this species.

I traded for one specimen consisting of thorax and pygidium intact carefully adding a

head of matching proportion from my collection. This was done as a reconstruction effort to mainly demonstrate how Pterygometapus appeared when articulated. The odds of ever finding a complete specimen are about zero to none. However someplace buried deep in the Martinsburg it could be possible many exist from an intact paleo-environment where they occurred abundantly.

Looking at this species simulates an ideal proto-type for species appearing later, such as Dalmanites, or Synphoria Stomata. These Silurian and early Devonian trilobites look very similar in appearance but the Ordovician Pterygometapus is found to be much smaller as with Flexicalymene compared to Calymene from Silurian rock beds. Heads look like typical Devonian age trilobites with large, distinct, raised kidney bean shaped compound eyes. The detailed glabella looks like familiar types from trilobite species such as Greenops. The body has segments similar to the average common type trilobite. The pygidium is triangular and somewhat pointed with a sloping arrangement of furrows distinguishable form the axel spine and outer plurons. The thorax appears to have about 10 segments and the pygidium has 13 divisions appearing more fused as a singular unit. The plurons have squared off broadened tips.

The most rare trilobites of Swatara Gap

Trilobite Mesotaphraspis Parva:

{number eight} This is certainly the smallest trilobite I have ever seen. At best any larger than the end of an eraser head on a pencil would be considered unusually large because they are very tiny and at times almost microscopic. The appearance is that of a typical trilobite. They look similar to the Devonian age Dechenella. From Swatara, these animals are associated with living environments in the paleo-world. They are extremely rare, occasionally found as isolated specimens. However they are not so difficult to find once certain rare paleo-environments are encountered in tact. If you are lucky enough to dig into the most unusual beds, suddenly you will often find a higher count for these rarities. This is an environmentally conducive situation. Certain delicate species are often found congregating in groups. That is the way this works. These trilobites usually needed a group of host species of other animals for protection and support. Being so small they are normally found together with groups of starfish and crinoids. I observed one specimen found by a man from Va. where several of these trilobites were facing each other in a small circle. Their mode of life was to exist in groups within busy habitats where the food supply was good and protective coverage was in their favor.

Mesotaphraspis had a smooth even surfaced crainidium and a larger proportioned featureless glabella with what appears to be compound eyes. It has 7 thorax segments and a brachiopod-like scalloped pygidium often exhibiting a centrally positioned talson spine. I have specimens of the starfish Phragmactis where there are several of these trilobites showing on the single slab together with small delicate crinoids.

The extremely rare Trilobites of Swatara Gap.

Trilobite Platylichas:

{Number nine} Here is another extremely rare species. It mostly seems to occur in sandy hash beds with debris apparently washed in by currents from other environments where it flourished. I was very lucky to collect my first and only complete (somewhat distorted) specimen from dump piles deposited north of the original site.

It is not much over a half inch and (probably never grew much larger) seems to exhibit 12 thorax segments with a broad pygidium having flared tips along the edge. The interior surface appears ornamented with dot-like patterns. The cephalon displays a squared off snout-like bill and eyes appear to be compound. The general appearance of Platylichas is like an average looking trilobite but more detailed and sculpted around its boarders.

Actual size of trilobite approx:1/2”

The most rare trilobites of Swatara Gap

Trilobite, Proetus Billings:

{Number 10 and most rare of all} I have only ever heard of 2 or 3 specimens ever being found. They are very similar in appearance to Mesotaphraspis but normally larger comparing adults of each species. I doubt if they grow much bigger than one half an inch. They appear similar to the Devonian age Dechenella having a typical appearance of a common trilobite. There is nothing much unusual about its taxonomy. Due to scarcity its environment is uncertain. Specimens I have seen are in coarser dark rock. I purchased a negative (counter part) of a specimen from another enthusiast who found it after a different collector from the region found the positive about a week earlier. Good luck has made it possible for me to complete the trilobite fauna for Swatara Gap.

The specimen appears to have at least 12 segments in the thorax merging with continuous segments in the pygidium. It is difficult to distinguish where the thorax and pygidium actually separate. The plural surface has a subtle beaded looking ornamentation. The cephalon is nearly identical in description to Mesotaphraspis.

In conclusion: there were speculative examples of an Agnostus variety of trilobite found however from photographs of a suspect specimen it seems difficult to interpret if that's what it really is for sure? If verified it would make the species count for trilobites even higher. That type seems to work better as a late Cambrian species.

Echinoderms from Swatara Gap, (Steleroidea)

The most interesting fossil to collectors from all levels are the starfish. From the times of my earliest visits it was the ultimate goal to find one. I was unaware of how many species were available to collect until one day I became exited about a find not representing what I thought would be the type to encounter. This looked like a small common type modern Atlantic star. The kind you might find washed up on the beach. Brittle stars were the only type I was aware of. There were a few named species of those types classed as ophuroids reported and listed in publications. The specimen I found was rigid and stellar from a group called stelleroids. I was soon to become aware of at least eight different species possible to find at Swatara Gap. I will outline the order of rarity and describe each type.

Taeniaster with two carpoids

Echinoderms from Swatara Gap, (Steleroidea)

Ophuroid Taeniaster Spinosis:

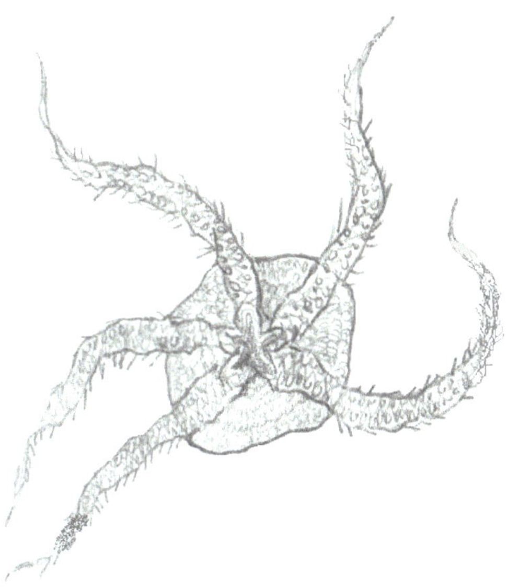

Considering many hours working rock at Swatara I never had a great amount of success finding exceptional specimens of this fantastic looking brittle star.

In later years near the top of the lower quarry in the second quarter from the south side there seemed to be a layer producing them to some degree.

I recall on two different occasions finding lenses in this shaley bedding where there were groups of this species to be encountered on slabs. My theory on starfish habitat was similar to a shallow bay area where living animals accumulate in small depressions on the floor. I have seen this in modern settings while searching for clams in the bay. Starfish, crinoids, and small trilobite species appeared to congregate in the same fashion. My earliest finds were not of the best preservation, small, and usually not very well articulated. In later years I encountered them in the same beds finding multiple specimens but again, nothing overly spectacular.

The best I have ever seen was one reported being found by a man along the railroad bed while hiking along Swatara Creek. It was presented to the local Museum where it was placed on loan for casting. The specimen was as perfect and large an example as could be expected. The oval body was the exact dimensions of a quarter (1 inch diameter) with neatly spaced 1-1/2 inch arms showing spines along the boarders and surface with intricate mouth details included. In later years while digging near the top on a bench in the second quarter another collector split open an excellent example, which was quite similar. The details were less pronounced in the central portion of the body but the arms were well displayed. Additionally there were rare carpoids included on both sides of the starfish. I expressed a deep interest marveling at the nature of this find. The man understood and elected to give me the specimen. Showing gratitude I presented his accompanying young son (who was the one interested in fossils) a complete cryptolithus to prepare at home (explaining how). Continued visits only ever produced minor examples of this rare fossil for me. I do recall finding both a Taeniaster and an Acidaspis within minutes apart on the forth quarter (north) side years ago. That starfish showed a textured dorsal surface on the body but the arms were not complete in length. A veteran collector later informed me better specimens often occurred in blocky sandy layers. Most all environments were suitable for this species. Living examples could be found among washed in debris in hash zones. Starfish were predators and their quarry was small mollusks, which were quite plentiful. One of my most interesting (one-of-a-kind) cabinet specimens is a silver dollar size shell (Rafinesquina) with a Taeniaster centered on the top. This could be an example of feeding but most likely the starfish was taking cover under the shell, to later become superimposed during fossilization.

Echinoderms from Swatara Gap, Steleroidea

Starfish, Ophuroid Protasterina:

Quite honestly I can only assume many specimens of small delicate ophuroids to be of this species. Their appearance is similar but the arms seem to be lacking the many spines Teaniaster has.
They appear to some degree, more like a modern serpent star.

I have collected various examples one of which seems to have a body appearing less oval in shape with long spindly arms in a sandy textured rock. Other examples are quite tiny. It may be possible juvenile specimens of Teaniaster could appear mistakenly similar to those thought to be Protasterina. In paleontology species are constantly being re-studied, re-described, and re-named. The nature and environment is most likely the same as with Teaniaster. The Ophuroid types were more often found as isolated specimens. They were able to roam the sea floor rather swiftly compared to the stelleroid types by use of their serpentine arms propelling themselves versus the use of tubular feet.

Echinoderms from Swatara Gap, Steleroidea

Starfish, Asteroid Mesopaleaster:

In my introduction I described finding this species in blocky sandy layers where most I am aware of usually occurred. My first specimen (pictured left) was not perfectly laid out in a perfect star shape.

I found others later in my collecting years, the best coming from dump piles north from the original site. They are usually the size of a dime to just beyond a nickel in full span. The arms are relatively thick and the entire body surface is porous or pitted. At times distorted examples could be found in a burrowing position. They seemed to be associated with hash debris and probably used those kinds of environments to their benefit for protection and feeding habitat.

I suppose they were mostly near shore predators and scavengers.

Echinoderms from Swatara Gap, Steleroidea

Starfish, Asteroid Phragmactis:

This species is pronounced as if it were spelled (Frag-mack-tis). I first became aware of Phragmactis while observing a slab collected by a staff member of the local museum. He proudly displayed his find exhibiting a fair number of individual specimens.

Along with the starfish were several tiny specimens of the trilobite Mesotaphraspis. The slab represented an intact paleo-environment where everything was alive upon burial. In later years I encountered bedding where slabs of the same kind appeared (quite possibly the same strike in the vertical bedding). The area was more or less near the central division of the quarry in a green shaley layer. I was able to pull out large plates by first removing the brick-like sandy layers inward and adjacent to this bedding in order to yield greater surface volume of the starfish bed exposed in the rock. I often used this technique to enter adjacent bedding planes with greater ease and less corruption of the section I was attempting to extract. A lot of patience and preparation are part of the discipline in field paleontology.

Hasty excavation can do more damage than one might realize. I was pleased with the volume and number of large pieces I was able to remove as far into the bench as I could go in one day. I was able to reassemble at least three slabs of positive and negative planes. I later received reports of more plates emerging (presumably from the same spot) about a week after by another local collector.

Phragmactis were again more star-shaped with thicker arms but not quite as rigid as Mesopalaeaster. Many examples portrayed flexibility in their arms. The bodies were modeled throughout the surface and covered in many spines.

These types seemed to reach an average span of around an inch and a half or about silver dollar size. I have seen them both in green shaley layers and also in sandier bedding always clustered in groups. There are several good specimens of about a dozen on one of my slabs. A few trilobites Mesotaphraspis also appear on these slabs I collected along with what appears to be partial examples of Teaniaster and several wispy, delicate, crinoids (Heterocrinus). This species, seemingly living in groups within intact Paleo-environments probably accumulated in shallow depression pools on the gentle muddy sea floor. Some I recently acquired from a fellow collector display in brownish sandy bedding appearing more like they were originally on a muddy bottom and covered by current swept sands and silt. I presume they were generally adapted to deeper calm environments. They were a type of starfish probably moving about slowly on tube feet well protected from predators even while exposed to the open sea by the covering of spines over their entire surface.

Echinoderms from Swatara Gap, Steleroidea

Starfish, Asteroid Petraster Prici:

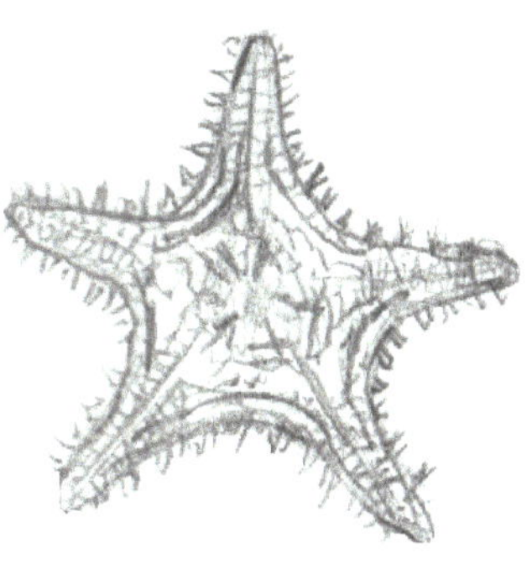

I first observed fragments of this species at the local museum where the gentlemen whom it was named for was a resident paleontologist. I am not sure whether or not any complete ones were ever found by any one else. One day I was in the (fourth quarter section) working bedding from south to north extending to the limits of fossil production. Splitting a pale smooth grayish piece of shale I saw the specimen flash in front of my eyes with a yell of excitement. This was extremely exciting to find a large (1-1/2 inch, probably average size for this species) stellar (Christmas cookie shaped) starfish having a shape more like the modern Bahama type star. One where the rigid star shape was such that the arms were wide from the attachment to the body tapering into a triangular shape. The body shows extensive surface detail and the entire parameter displays protective spines. I cannot be certain of the nature of this species since very few seem to have been recorded. It was probably a bottom feeding deep-water creature moving about slowly with tubular feet, protected with sharp spines to ward off predators. I would expect these rare types might have occasionally surfaced near the contact of the Silurian conglomerate from the less prolific, adjacent shaley layers.

Echinoderms from Swatara Gap, Steleroidea

Starfish, Asteroid: Salterella Ulrich bellulus:

I have never encountered a single trace of this very rare species in the field but was lucky to be gifted a negative half containing two examples. The nature of this particular species seems like a cross between a spindly-armed type (like a Ophuroid) and Mesopalaeaster. The arms have many pits like Mesopalaeaster covering the body but it has no disk. This starfish like an Ophuroid appears to have been able to use its serpentine-like arms for rapid movement across the sea floor. It also was lined with spines for protection and probably supported itself well in various environments. My multiple specimens occur in coarser brown sandy material.

Echinoderms from Swatara Gap, Steleroidea

Starfish, Asteroid: Schuchertia:

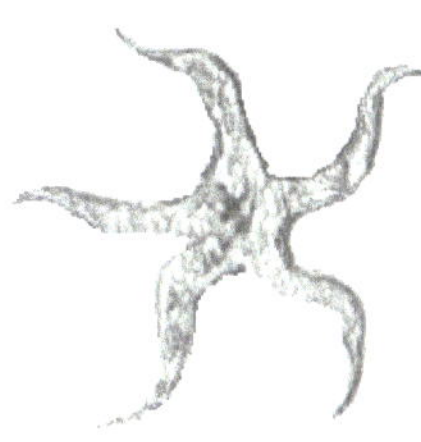

Apparently an extremely rare species my first awareness of this type was from observing specimens in museum archives. I was able to make a sketch of a specimen for reference. I was never aware of any of these showing up in the field neither by myself, or any collectors that I knew. I describe it as a spindly-armed stelleroid where the arms are thicker near the body and become tapered as they flare outward. The entire body appeared generally smooth but pitted throughout. I could not make out any visible spines. The only example I have to show is a created model simulating what a specimen might have looked like. I imagine these could have grown to a maximum size of a half dollar. It was another species where the spindly arms allowed it to propel itself having better mobility using its arms to supersede use of tubular feet. The rare types as described throughout this report might have been more prolific out of the boundaries of the immediate paleo-environment. I call these drift-in specimens. I tend to imagine somewhere in the formation it could be possible larger numbers of such unusual specimens could exist in any category.

Echinoderms from Swatara Gap, Steleroidea

Starfish, Promopaleaster:

Apparently the most rare of all and the giant starfish from Swatara is what I describe as the platy or turtle shell stelleroid. I recall seeing fragments of specimens in museum archives. Its shape is like that of Petraster but the size apparently reaches the four-inch mark and perhaps larger. It could be likely this species was easily subjective to disarticulation and poor chances of preservation because of the large size. I saw one example collected by a local enthusiast where it displayed mostly the outline of the body. It appeared as though a trilobite (or whatever) chewed its way directly through the center of the specimen virtually destroying all of the surface detail. The odds of finding a complete articulated specimen seemed zero to none. I imagine this type with armor like plating would lie on deeper calm sea bottoms. It was probably sluggish and moved about on tubular feet. The armor presumably served as protection in those kinds of environments.

Stelleroid: Promopaleaster

Echinoderms from Swatara Gap

Crinoids: Second to starfish but of great interest to the paleontology of Swatara Gap were various delicate species of crinoids to be found. Calm paleo-environments were apparently well suited for them since articulated finds were not uncommon. I often referred to Swatara's Martinsburg as Pennsylvania's Ordovician answer to the Cambrian Burgess shale's of British Columbia Canada. I promote this concept simply because of the excellent preservation of many fine delicate structures. I have collected many nice examples with column, calyx, and ambulacra all in tact. On a few occasions I found feathery ambulacra with attached pinnules. Some were evenly spread on a rock where there were five pairs of dual branches making ten in all forming a circle from the mouthpart. I will attempt to describe a few different examples.

Crinoid, Ectinocrinus simplex:

I have encountered these on more than one occasion. They are often folded into an umbrella-like pod with thin column attached. When closed they sort of remind you of a hibiscus bud. The calyx and ambulacra seem rigid, segmented, and not so feather-like. I found them in sandy-bedded layers and in better condition from greener shale's. The calyx was usually closed in the sandy layers indicating they could have been swept in from more subtle environments probably deposited by turbulent waters during storms. The variation in rock texture suggests periodic turbulence. The clastic difference between layers and the average distances in separation might reflect patterns in stormy turbulent periods indicated throughout the lithographic bedding profile.

Echinoderms from Swatara Gap

Crinoid, Glyptocrinus:

I found these specimens as very delicate and small (around an inch and a half at the most). Good complete examples are often curled and were most likely buried in life. They were found in lighter gray shale's and probably were from an intact, deeper, calm water paleo-environment. These show intricate details in the segmented multi-branching ambulacra with feather-like pinnial structures. The columns have distinct spines on each segment. The spines seemed to be a component to the nature of many organisms from the echinoderm group suggesting open calm waters, thus a need for external protection to intimidate predators.

(34)

Echinoderms from Swatara Gap

Crinoid, Heterocrinus:

Another very delicate species often found in groups with slabs of the starfish Phragmactis. These are mostly wispy broom like in appearance and look more like filamentary stems and branches with indistinct calyx. They needed calm enriched environments for their success among the neighbors within their delicate habitat.

Echinoderms from Swatara Gap

Circo-crinoid Echinosphaerites:

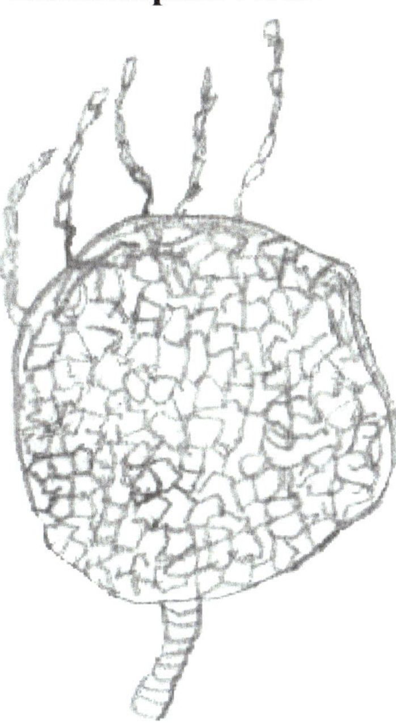

I'm sure there are other species to be identified from various finds to extend a detailed list. One in particular is a specimen I found in green shale's from the middle of the first and second quarter on the south side of our formation. It is a spherical (bulbous) cystoid-like calyx with a surface ornamentation of somewhat hexagonal shaped plating showing a few portions of ambulacra extending from the upper mouth region and a partial column attached. The column appears more uniform rather than segmented exhibiting no spines. I am only aware of two specimens ever being found. From the nature of the short attached column the specimen may have drifted into the local environment by currents. Animals of this type probably gathered abundantly in colonies from other more suitable places in the paleo-environment within the Martinsburg.

Brachiopods from Swatara Gap

Brachiopods are one of the most common types of mollusks to be found in the fossil record. There are claimed to be well over the ten species I portray, found at Swatara. I encountered so many I often tossed them over the piles only saving those unusually aesthetic or exceptionally interesting. In the first quarter there seemed to be many hash (sandy layers) filled with both brachiopods and pelecypods. There used to be a gully-wash near the top behind the southern most bridge abutment as well as other exposed sections in the southern side of the quarry excellent for multiple hash slabs including trilobite parts. Brachiopods are shellfish described as being attached by a stalk raising the animal above the surface of the sea floor to filter feed. They can be separated from pelecypods (clams) since they are bi-symmetrical. A brachiopod when divided from the narrow hinge part through its center will appear the same on both sides whereas plecyopods are not bisymmetrical. Plecyopods are burrowing mollusks navigating the muddy sea bottom with a pseudo foot used for mobility.

Brachiopod, Rafinesquina Alternata:

The large Rafinesquina is one of the most common and interesting specimens to collect at Swatara gap. This signature fossil predominates many Ordovician localities often preserved in Pa. limestone deposits. The shell usually extending beyond semicircular often grew to an inch and a half. They were often quite aesthetic usually maintaining a bright orange oxide coating in contrast to the dark matrix. These were usually common both in sandy beds and greener shale's. The surface usually exhibited mild striations from the hinge radiating to the edge of the shell. There was a prolific bed in the second quarter yielding specimens in quantity, often producing very nice multiples.

Brachiopods from Swatara Gap

Brachiopod, Leptaena Gibbosa:

Very commonly encountered throughout the bedding in the Swatara Gap fossil unit. They usually appeared with raised relief and were nut-like in appearance with striations radiating through the radius of the shell. They were numerous in bedding near starfish layers in the second quarter. This species was common in both sandy beds and green shale's.

Brachiopod, Dalmanella Multisecta:

This species: (usually not much larger than a dime in size) aesthetically pleasing to the eye is commonly found together with other fine specimens of echinoderms and trilobites. It is a circular shaped mildly scalloped species occasionally found with both valves still attached. They seemed common throughout the fossil unit but more numerous in sandy beds.

Brachiopod, Sowerbyella:

Another signature species, it is very common in many Ordovician rocks throughout the region. It is so common and plain looking most times I used to ignore these when found. They are pretty much a featureless narrow semicircle shape. There is a bed in the third quarter packed with this species running the entire vertical length through the formation. I often used this layer as a reference point to being near a prolific trilobite area. The species seems very common in the green shale's but found throughout the fossil unit.

Brachiopod, Sericoidea Plicatella:

This species is tiny and mildly scalloped. It appears about 2-3 mm in size usually found in clusters. They can be found throughout the formation most often in greener shale beds. It is abundant in bedding from the second quarter where Cryptolythus and Triarthus are prolific.

Leptaena Gibbosa

Dalmanella Multisecta

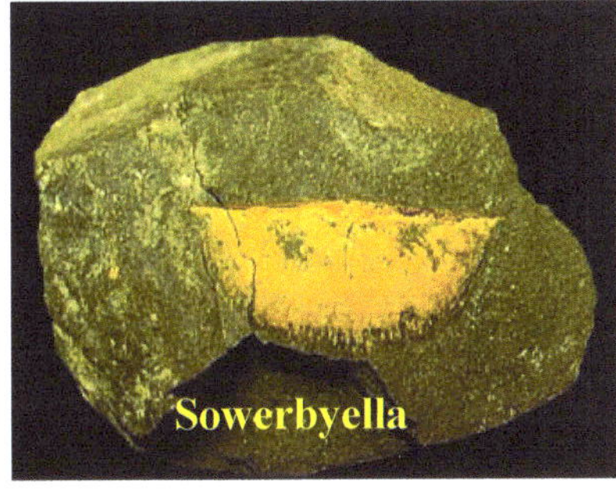

Sowerbyella

Sericoidea Plicatella

10X

Brachiopods from Swatara Gap

Brachiopod, Hebertella:

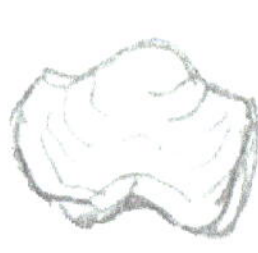

This species with surface appearance barely distinguishable from a pelecypod is not very common. It has more of a heart shaped appearance with a smoother surface and growth rings. It seems to occur mostly in greener shale beds when encountered.

Brachiopod, Lingula:

This is an example of a living relic noted for being the oldest surviving brachiopod. It is fairly common throughout the fossil unit at Swatara easily recognized by its elongated oval shape. Generally, for whatever reason, these normally were found black. They usually do not exhibit an iron oxide coating. The surface appearance is rather dull usually bearing visible growth rings appearing somewhat like a modern mussel. The average size is about ¾ inch.

Brachiopod, Craniops:

Appears very much like a Pholidops but is usually very tiny showing vivid growth rings. The size is usually around 2-3 mm.

Brachiopod, Pholidops Subtruncta:

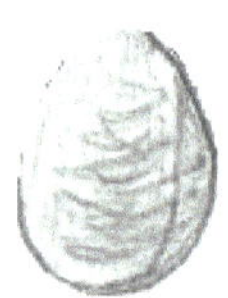

This brachiopod looks very similar to Lingula but is usually more round in appearance. (Less elongated). Most specimens are also uncoated with orange oxide looking similar to a pelecypod usually preserved with a paler patina surface than Lingula. These are found in shaley beds throughout the fossil unit.

Brachiopod, Plectorthis Plicatella:

Appears very much like Dalmanella in shape and size. Heavily scalloped in appearance. It seems more common in the sandy hash beds.

Hebertella

Lingula

Pholidops Subtruncta

Craniops 30X

Plectorthis Plicatella

Pelecypods of Swatara Gap

As a fossil collector, I salute those in the academic profession who endure the patience of studying the various clams occurring in the fossiliferous rocks of Pa. Personally (with no insult intended) I find the mollusks, particularly clams, to be rather dull and somewhat boring. I have encountered many specimens from Swatara often tossing them with little regard (and often do the same at many other sites). However in the event of compiling a fauna assemblage for a site, I do make it a purpose to collect a representation of the best examples possible in order to create a thorough suite.

If one does take the time to examine and study these it can be interesting to see many differences various species exhibit in there structures. Pelecypods can at times be more troublesome to correctly identify nevertheless there are a few larger and more distinguished forms to portray and attempt to identify from Swatara Gap. Many are rather minute in size. Some of the larger ones can reach the size of a modern mussel, perhaps around an inch and a quarter. Finding Pelecypods puts one in bedding highly conducive for them since they were subsurface dwellers for the most part. They seem common in rather barren layers, a habitat of their life mode as burrowers. I will exhibit many examples from my collection excluding detailed descriptions as to the nature of each individual.

Bedding to the south in the first quarter around silt beds often produced slabs containing multiple pelecypods. Plecyopods usually did not take the orange oxide typical in most of the Swatara Gap fauna. These were usually preserved with a silvery-black shiny surface. Perhaps occurring deeper from bedding parting induced less oxidization.

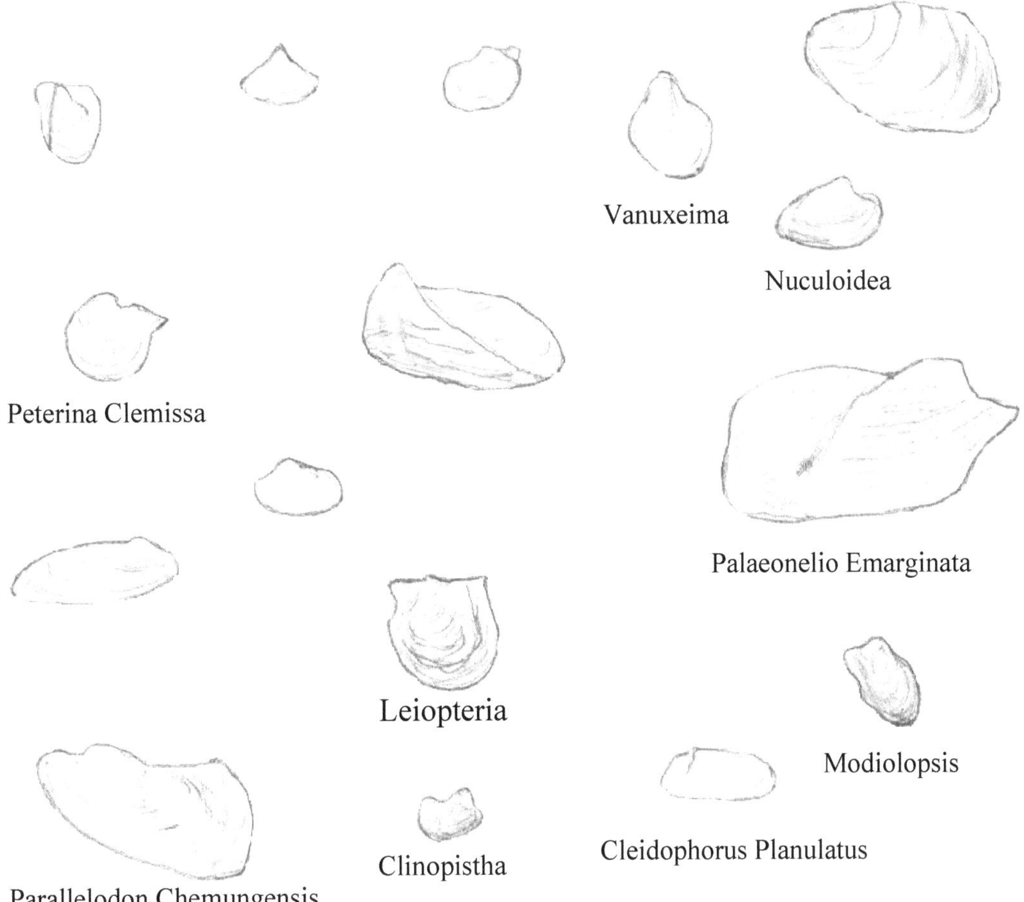

Pelecypods of Swatara Gap

Gastropods of Swatara Gap

Sea snails were fairly common to encounter throughout the formation at Swatara. Most were usually small (about the size of a dime) there were only a few major species and others I found might extend the list.

Common Gastropod, Liospira Micula:

Very common spiral forms seen throughout most of the formation. The slatey shale's near the center of the quarry produced frequent encounters of this species

Common Gastropod, Sinuites Cancellatus:

Globe-like, subtly coiled they generally lacked the signature orange oxide usually taking on a silvery hematite appearance. These could be found in green shale's and sandy beds.

Uncommon Gastropod, Lophospira Ohioensis:

This was a type having more of an extended shell growth (conch-like) infrequently encountered.

Rare Gastropod, Hormotoma? This tiny shell extends outward in growth.

No identification Coiled and chambered looking

Sinuites

Sinuites Cancellatus

Liospira Micula

Lophospira Ohioensis

Hormotoma

No identification

Common specimens from Swatara Gap

Bryozoan's :

These are branching sponge-like animals growing from calm sea bottom environments. They are fairly common to find in the Martinsburg shale's at the Gap. Many times I ignored these when encountered as most were disarticulated or fragmented. Later in my collecting experience I captured a few better examples before the site closed.

Bryozoa, Rhinidictya:

Appears with detailed surface structure, thin and branching.

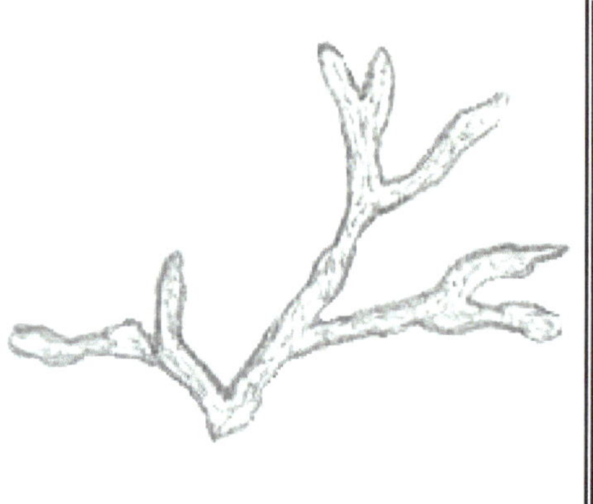

Bryozoa, Dekayella:

Found as spongy branching specimens with a porous surface.

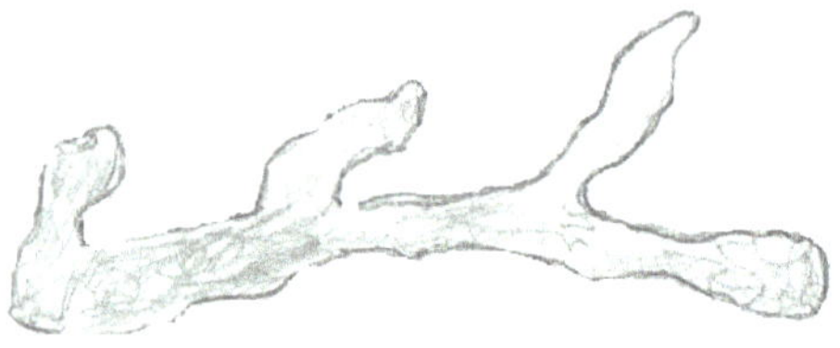

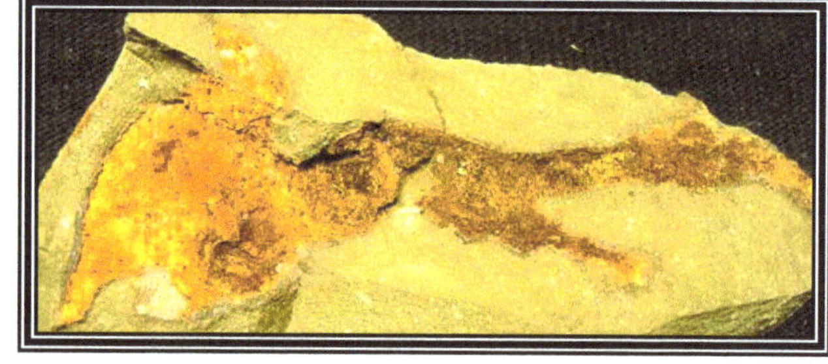

Crustacea, Ostrocods:

Often overlooked are many microfossils. Tiny carapaces from the remains of free-swimming barnacle-like arthropods were common in the late Ordovician and Silurian seas found in fossil marine units throughout Pa.

There are apparently several species found in the rocks from Swatara gap. They are nearly microscopic appearing as oddly shaped half-moon-like casts.

Common specimens from Swatara Gap

Annelids, (Marine Worms):

Trace fossil worm burrow

Trace fossils from marine animals can be found in the Martinsburg shale's. Among such features as trilobite burrows and occasional trails are castings of burrowing tubeworms. Soft-bodied animals of these sorts probably existed along with other similar creatures. Worms probably were types with feather like arms for filter feeding similar to modern marine polychaete types. Other soft fleshy examples rarely fossilized could have been marine plants. (algae-like forms). Bits of carbon are often found which may be all that remains of plants likely to have existed in shallow sunlit waters of an Ordovician sea.

Hydrozoa, Graptolite Diplograptus:

Usually a small carbon-like fossil or filamentary remains having well defined detail. They are long thin and barbed-like in appearance. These were thought to be the hard parts of an animal from a massive floating colony similar to jellyfish. Grayish shaley beds far south in the first quarter of the fossil unit at Swatara seem to be rather barren with the exception of an abundance of these specimens. They are also occasionally found scattered throughout other sediments.

Graptolite

Cephalopod, Michelinoceras:

These are so common at Swatara, dare I say, almost to the point of being annoying. The conical shell casts are found throughout the formation. The second quarter of the formation has beds where they were quite plentiful. Most times they are found in the shaley bedding suggesting these animals lived in deeper water. This predator, which was much like a squid with an elongated conical shell, grew to very large sizes in Ordovician paleo-environments. Presumably, examples exhibiting variations suggest a few different co-existing species were prevalent.

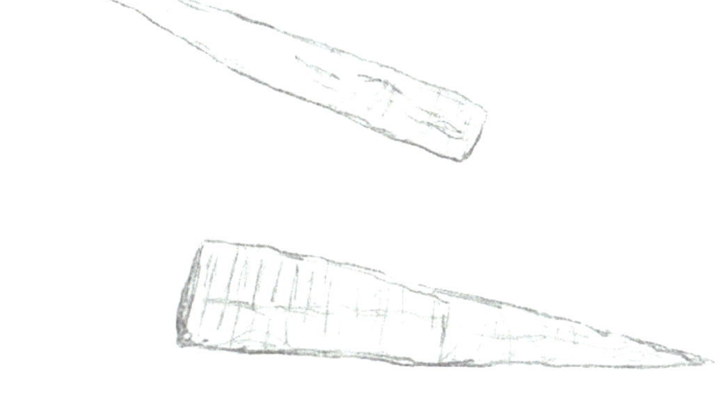

Michelinoceras

Rare specimens from Swatara Gap

Carpoids:

A most peculiar specimen turned up on a shale split. The creature looked like half clam and half sea horse. This was an extremely rare specimen of a carpoid. The tail part appeared to be capable of grasping and attaching to objects while the head part looked like a split bean or peanut shaped affair. It was found in green shale. The next time I saw them did not come until a specimen (previously mentioned) of Teaniaster showed up with two small examples on either side of the slab. This indicated an association within a living paleo-environment where extremely rare specimens can turn up in multiples when found in place.

Conularia, Trentonensis:

I recall a day while visiting the site another individual approached me with an unusual specimen to identify. I examined this unusual whitish triangular shaped object with subtle designs scored on the surface. During this early stage of my experience I was unfamiliar with anything like it but suggested it was of interest and worth studying. The man decided to donate it to me for research. Later I learned it was another very rare encounter. This was from a different class. Conularians are four-sided skeletal frameworks tapering to a cone-like structure. In theory the pointed end was planted in the sea bottom and soft tentacle-like tissue apparently protruded from the aperture of the shelly body. Another idea was it might have been free swimming in the theoretical sense applied to graptolite colonies (jellyfish-like in its mode of life). I later recovered partial specimens showing more detail on the surface exhibiting an enhanced groove-like structure.

Rare specimens from Swatara Gap

Lichenids:

I found peculiar elongated body structures having four spiky protrusions. This aroused curiosity as to weather these were partial crinoids or perhaps disarticulated starfish. I attempted to research them based on the general appearance noticing some resemblance to Lichenids. These animals could occur in Ordovician environments. One is situated on a slab with a crinoid though their true nature is speculative at best.

Arthropods, Eurypterids:

To the best of my knowledge there are many reported examples for evidence of these creatures potentially showing up mainly as skin fragments. The large water scorpions predominated marine waters mostly during the Silurian period continuing back into the late Ordovician. Many species grew to very large sizes. These were fast free-swimming, probably deeper water predators perhaps too large to hold up for complete preservation. No complete or even good partial specimens were found at Swatara Gap that I was ever aware of but there are suspect fragments of potential remains.

Conoidal shells:

Small elongated shells of other animals also can be observed in the micro realm from the Martinsburg exhibiting a similar appearance to Tentaculites, which is common from Silurian rocks in other formations throughout Pa..

Unusual specimens from Swatara Gap

Starfish Phragmactis

Starfish Teaniaster superimposed on brachiopod Rafinesquina

Cluster of Rafinesquina

Unidentified Plant or animal?

Starfish, Phragmactis

Unusual specimens and uncertain affinities from Swatara Gap

Ordovician Fossil Locality in Bellefonte, Centre County

A well-known locality along Pa. Route 144 is mostly good for collecting abundant examples of the trilobite, *Triarthus Becki.* This site occurs in the Antes formation, adjacent to, and in contact with, the Coburn limestone formation. The trilobites occur in a chocolate brown shaley rock that splits, showing bedding planes often completely filled with disarticulated examples of Triarthus. This trilobite usually appears at a maximum of an inch and a quarter in length, if an articulated adult is collected within this softer section of paleo-lithography.

One point of interest was the coloration of the fossils. The trilobites at this locality are all coated with a thin white mineral, giving them an aesthetic contrast for fine exhibit material. Complete specimens are infrequent, but probability will allow complete specimens, with patience, and a lot of exploration within the rich bedding. One can be content to find slabs filled with heads of this trilobite, if only searching for samples. Headless bodies will be found, as well, exhibiting thorax, pygidium and axial spine.

Devonian age marine fossils from the Mahantango Formation

From the famous Swatara Gap fossil site one could always make a trip an additional six miles to the north moving forward in time to collect Devonian age material.
The site, just north of the Lebanon-Schuylkill County border always gave the opportunity for a pleasurable ride through the Swatara State Forest on a narrow rugged road following Swatara Creek. Traveling through the cool evergreen groves, made for a nice lunch break after a full morning of collecting at the gap. One could always take the speedier and better-paved Rt. 72 then turn on Swope Valley Rd. arriving at the site by using an unpaved road and coming in from about a mile to the north. I always preferred the lumpier, more scenic route, which was just a matter of crossing the old steel bridge (presently being demolished for new construction) on the Swatara Creek at the gap, making a left turn, and heading upstream. Once there, you came upon an open face of rock outcrop to the right. This site was always pleasurable for it's spacious quiet outback environment, surrounded by cool streams and towering evergreens. Back in the 1970's and 80's, there was still a cliff-like embankment where one could easily access quality slabs of fossil reef material. The site known as the Suedeburg locality is part of the Swatara State Park. I always found the richest beds in the southernmost part of the exposure best for significant finds. It wasn't uncommon for a busload of students to arrive making use of this site. Most interesting, were sections producing large slabs completely filled with colorful iron oxide stained brachiopods and pelecypods, as well as casts from corals and bryozoans in a rich greenish layer crosscutting the ridge at a sharp angle. There were numerous well-preserved trilobites and partials in the same layer. I searched for exceptionally well-preserved specimens of the trilobite species *Dechenella* usually laying flat within these beds. Other trilobites, such as *Phacops,* were commonly found in less than satisfactory condition from adjacent layers.

To the south of the green layers were pale bedding layers where fossils were scattered within. These were usually white stained small trilobites from the *greenops* group found with abundant pelecypods. Most of the northern part of this site offers a scattered assortment of brachiopods, but has yielded very little to make collecting interesting. Today, the walls of the site are smoothed down, appearing to be both over collected and damaged by recreational vehicle activity. At present, collecting the once prolific layers poses difficulty, unless a new cut could be made within the bedding of the enriched zone. Good fossil collecting was available for a short while when the parks decided to dump at least 20 piles of material on this site from the Martinsburg Formation during construction at the Swatara Gap site. The Ordovician material added to the interest for collectors, while it lasted.

Fossils From Suedeburg Site, Schuylkill County

Trilobite Greenops

Trilobite Dechenella

Trilobite Dechenella with brachiopods

Trilobite Dechenella

Slab of Brachiopods

Plecyopod Actinopteria

Devonian Age Fossils, Perry County…

Crossing the Susquehanna River into Perry County, near the village of Alinda, a small road cut on Indiana Road, produced an unusual occurrence of the Lower Devonian age trilobite, ***Odontocephalus Aegeria*** (pictured below) from an out crop of shale within the Onondaga Formation. Other species included trilobites *Leonaspis*, (similar to *Acidaspis* from Swatara Gap), and various species of *Phacops*. The site is no longer as functional and productive as in the past. I had a few brief visits and missed collecting there in the most prolific days. Much of the productive rock outcrop is currently depleted or submerged and rather difficult to access. Many excellent specimens were collected from better layers by veteran collectors.

(42)

Devonian Age Fossils, Perry County…

The middle Devonian age rocks of the Mahantango Formation throughout various locations in Perry County yielded an assemblage of excellent marine fossils. These were deposited in a sandy delta, in a sea that existed more than 380 million years ago. Various sites allowed excellent places for collecting brachiopods, pelecypods, corals, and trilobites, occurring from Newport to Ickesburg and beyond. Shale pits in the Newport area were well noted for fossil corals from ancient reefs, often producing specimens of original material left intact. The large trilobite, *Dipleura Dekayi,* often attaining lengths of up to ten or more inches was always well sought after by advanced collectors, occasionally showing up as complete specimens.

Many trilobites from these deposits appear as they were fossilized most likely while in a living position. Most times, the species *Phacops,* when encountered occurred in areas producing them in large numbers. Certain beds or zones produced specimens in an enrolled or slightly curled position, an indication of being buried by current-swept sediments, while trying to escape by curling into position for protection. Many of these beds also yield large specimens of brachiopods and pelecypods.

Horn Coral Heterophrentis

Devonian Age Fossil, Coral Perry County

Devonian Age Fossil's, Ickesburg, Perry County

Trilobites Phacops Rana

Trilobites Phacops Rana

Trilobite Dipleura Dekayi

Trilobite Neometacanthus

Brachiopod Mediospirifer

Brachiopod with trilobite Phacops

Devonian Age Fossils, Ickesburg, Perry County

Gastropod (snail)

Gastropod

Dipleura

Enrolled Trilobite Dipleura Dekayi

Brachiopod with Devonian age driftwood

Brachiopod

Devonian Age Fossils, Seven Stars, Juniata County…

An old favorite locality is a shale bank along Pa. 235, a peaceful backcountry road in Juniata County. On my earliest visit there, I found collecting to be good. Broken layers along the edge of the outcrop were from a prolific zone. This locality seemed to me, to be the one that might yield a decent, large specimen of the trilobite, *Dipleura Dekayi*. On my first visit (early 1980's) I was finding many heads, pygidium, and thorax parts. Based on regular check-ups the shale pit seemed inactive for decades. Recently, {2008} the pit was worked again for fill material and large, excellent quality, complete specimens of *Dipleura* were found. I also discovered blocks of shale containing complete juvenile specimens. Of special note, were fairly large, abundant, well-preserved gastropods. With this locality, the specific beds being worked are most important. The bedding profile however is difficult to follow since pelecypods and other burrowing organisms moved about through the sediment prior to their fossilization, having disturbed much of the distinguishable bedding.

Pelecypods are quite abundant and get rather large at this locality. One excellent example of a Plecyopod to be found at this site is the razor clam species, *Orthonota*. They appear similar to modern species living in local bay waters today. At times, certain beds will yield specimens of the trilobite, *Greenops Boothi*. I have never found any evidence of the trilobite, *Phacops*, at this site. Occasional fossil wood fragments are found mixed in with the marine fossils indicating a near shore environment. This locality lies within the Mahantango Formation resulting from a shallow sea existing more than 385 million years ago during the Middle Devonian Period.

Devonian Age Fossils, Seven Stars, Juniata County

Trilobite Dipleura Dekayi 2"

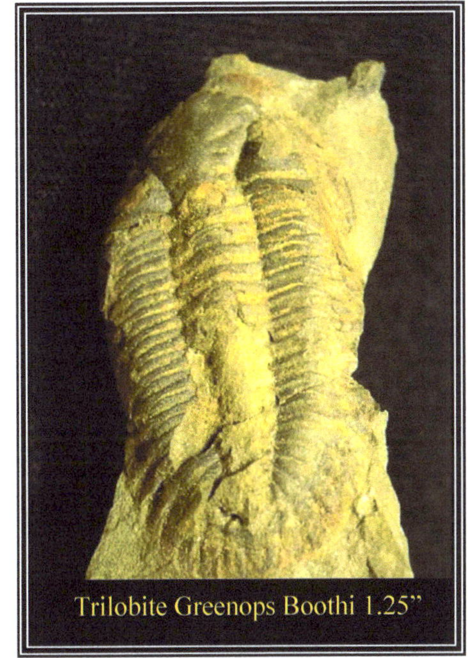
Trilobite Greenops Boothi 1.25"

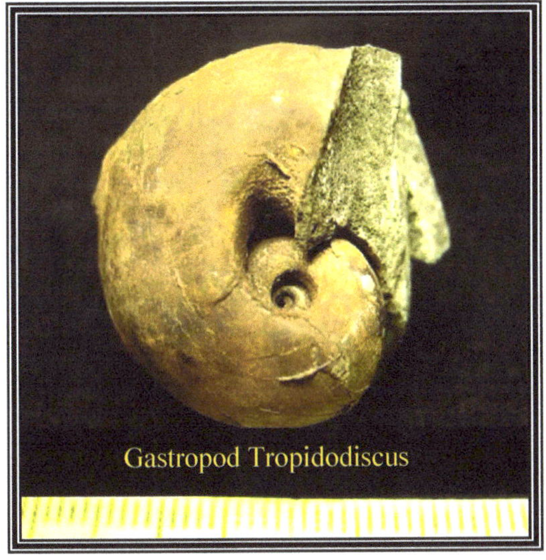
Gastropod Tropidodiscus

Pelecypod Orthonota (razor Clam)

Pelecypod (clam)

Brachiopod Mucrospirifer

Devonian Age Fossils, Seven Stars, Juniata County

April, 2008 8" trilobite Dipleura Dekayi with Greenops Boothi

Fossil driftwood

Cephalopod?

Devonian age fossil marine brachiopod slabs

(50)

Lower Devonian Occurrence of Rare Fossils, Centre Co.

During the 1970's, a site was discovered and worked by staff from our local museum in Lancaster, {where I had been a member of the Junior Earth Science Club}. The setting was east of Milesburg in the Lower Devonian Shriver Formation. Fossils were collected near an old mill site and in a local farm field. Remains of collapsed buildings produced fossiliferous punky yellow-orange and tan colored Shriver sandstone rock. Many superb specimens of rare trilobites, conularians, carpoids, and brachiopods were collected from loose block found in and around the boarders of cultivated fields for the compilation of a scientific collection for the museum.

In later years, a few of my mentors revealed the locality. I finally had a chance to experience what collecting was like there. The best workable area was along an old canal near an open field.

The site was noted for fairly large trilobites, *Synphoria Stemmata*, *Dalmanites*, *Trimerus Stelmophorus,* and the smaller *Leonaspis*. The actual outcrop used to extract rock for the buildings was not located to any specific degree. The Shriver formation does exist intact at various locations throughout the region. It is presumed material was quarried within the area.

Trilobite Synphoria Stemmata 8cm

Trilobite Trimerus Stelmophorus 17cm

Fresh Water Devonian Fossils: From Clinton County

An incredible site, currently famous for its world class assemblage of fossil, fish, tetrapods, arthropods, and plant remains, lies within the red rocks of the Catskill Formation deposited over 360 million years ago. This time, we are not looking through the usual marine sediments from an ancient seabed, but rather, a rare occurrence of an inland fresh water fossil environment. During the time these fossil animals were living, the picture showed ancient rivers cutting through the region, situated within a tropical setting. The region occasionally flooded depositing sediment carried from a great mountain range, situated to the east. Flooded channels along the river allowed pooling of water trapping a variety of aquatic species. Sediments buried these animals, which became fossilized in a low oxygen zone ideal for lack of decay and good preservation.

Found here, were bones, scales, and teeth, from ancient fresh water fish that often grew to be several feet in length. Also remains of plant life, and an occasional arthropod that flourished along the shoreline of the Devonian age forest. Studies are now under way on many species of early fishes and rare tetrapods identified from this site, lying along the west branch of the Susquehanna River. The more commonly recognized types of fish from this era were Placoderms, found as fossil bony plate remains within the bedding. There is a prolific layer producing the best fossils in this distinct reddish colored formation ranging to about six feet in depth. One section produces fish remains more so than another section in green shale along the same bed a few hundred feet east where plant remains are most abundant.

Of great importance with this site, known as Red Hill, near the village of Hyner, was the discovery of early tetrapods, such as *Hynerpeton Bassetti* and *Densignathis Rowei*. This site is one of great importance since it represents much of the first evidence of type animals emerging onto land as (tetrapods), four legged creatures, evolving from fish. The site is only collectable under special supervision. It is not generally open to free access by the public. This is an academic site worked carefully in order to preserve rare and valuable fossils for research. While having the privilege to collect there, I was able to accumulate excellent specimens representing aesthetic and scientific presentation of what Central Pa. was like inland from the sea during the Paleozoic era.

Fresh Water Devonian Fossils, From Clinton Co.

Placoderm trunk shield Groenlandaspis Pennsylvanica

Fish scale

Megalichthydid

Fish scale

Fish scale Hyneria Lindae

Dorsal fin spine to placoderm Turrisaspis Elektor

Fresh Water Devonian Fossils, From Clinton Co.

Hyneria Lindae teeth (large fish)

Tooth

Tooth

Hyneria Lindae 10x

Ageleodus Pectinatus (shark) 2mm

Hyneria Lindae tooth 2"

Tooth

Discovery of a rare bone from a Devonian age fish

Devonian Remains of an Acanthodian Fish, Gyracanthus

(56)

Shoulder Blades Teeth and Spines of Gyracanthus

Red Hill Fossil Clinton Co.

Fresh Water Devonian Fossils, from Clinton County

Fin spines of a large Acanthodian fish Gyracanthus

Devonian Age Fossil Plants, from Red Hill, Clinton County

Plant Fossils, Red Hill Site, *Archaeopteris*

Plant Fossils, Red Hill Site

Rare Plant Fossils, Red Hill Site

Fresh Water Devonian Fossils, from Clinton County

View facing toward the west

Fresh water Devonian fossils, from Clinton Co.

Skull material

Skull material

View towards the east

Rare new bone

Hyneria Lindae tooth

Unpacking, preparing, and studying fossils in the lab

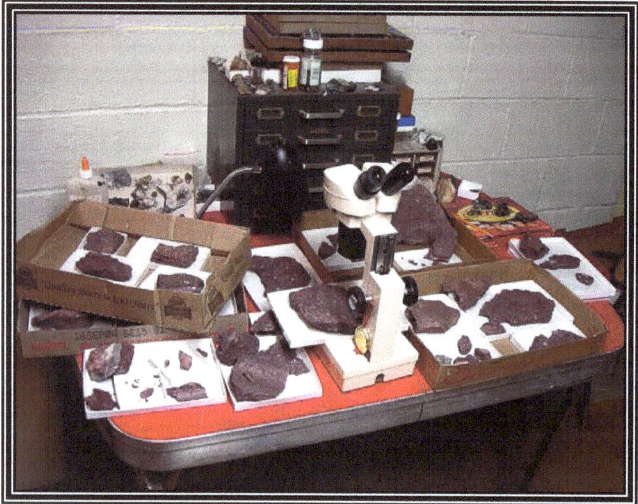

Plant Fossils from the Carboniferous Period

Fossil collecting for plants in Pennsylvania brings us into the Pennsylvanian period. The coal forming process began its history in the region over 300 million years ago, with constant flooding of swampy lands. Layers of sediment were left behind burying plant life creating an accumulation of carbon remains. Often found between many of the major coal seams and coarse rocky sedimentary rocks in Central Pa., are thin shaley layers yielding excellent plant fossil remains exhibiting superbly preserved detail. The Llewellyn formation of Schuylkill County, within coal bearing regions near Pottsville, produced numerous world class specimens noted for contrasting black shale, with fossils preserved in a white pyrophyllite (aluminum silicate).

Many years ago, an incredible site now lost and reclaimed as forest, produced some of the finest examples from the Pennsylvanian age in hard shale. Pinkish color matrix, exhibited fossils with excellent relief as a chocolate brown. I was able to obtain a few good samples from material previously collected (by my geologist friend mentioned in the introduction) from an elusive location in Schuylkill County.

I also had a few opportunities to collect nice plant fossil material from dump areas near coalfields at Broad Top, Huntingdon County. Arthropods and insects were reported from shale's of the Llewellyn formation but in all my years of collecting, I have seen very few of these. I found one specimen having taxonomy representing a cockroach. I heard many reports of dragonfly wings being found.

I once displayed a slab exhibiting small glass shrimp in a showcase, but somehow misplaced the specimen within my collection. The lesson here is the importance to mark such items for future reference, I hope the specimens are on one of the slabs hanging in the back wall of my main fossil case. The plant fossils exhibit many different species some of which are common and others rare. Various fossil tree trunks, leaves, roots, stems, and unusual seedpods are to be found. The best fossils are ones with spacious distribution on a slab, and the larger the slab one can extract, the better. I have seen clumped matted fossils, which rate low on the scale of how I evaluate a specimen. One should aim to collect hard material that splits very nicely with as many full, neatly separated leaves as can be found, showing contrasting matrix inbetween. This allows the most aesthetic appearance for display. Some of the layers will have a lot of fossils, but can be mealy looking and not very well preserved, those I generally avoid collecting.

Fossil Seed Ferns, Broad Top, Pa. Huntingdon County

Plant Fossils, Broad Top, Huntingdon County

Seed Pod

Seed fern

Plant Fossils, Schuylkill County

Annularia Stellata

Annularia Stellata

Pecopteri

Plant Fossils, Schuylkill County

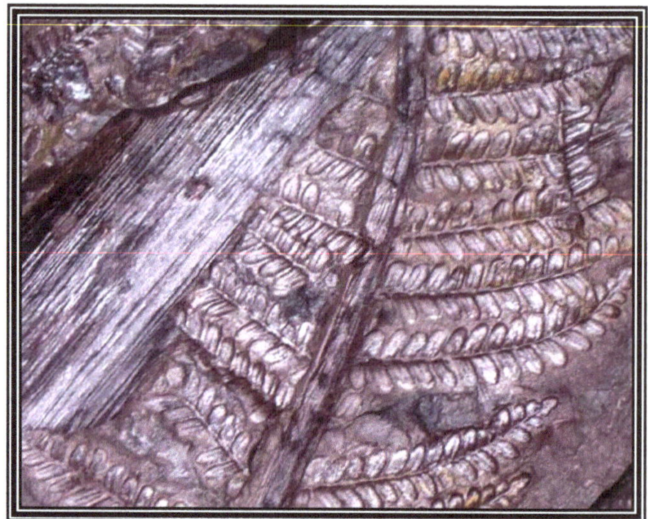

Neuropteris Scheuchzeri

Pecopteri

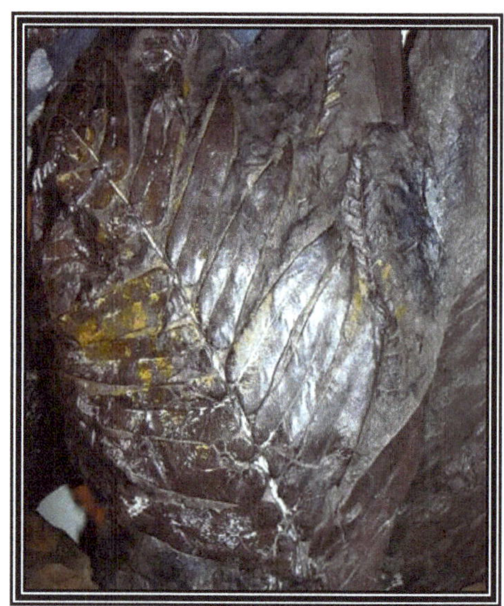

Plant Fossils, Llewellyn Formation, Schuylkill County

Assorted species

Sphenophyllum

Plant Fossils, Llewellyn Formation, Schuylkill County

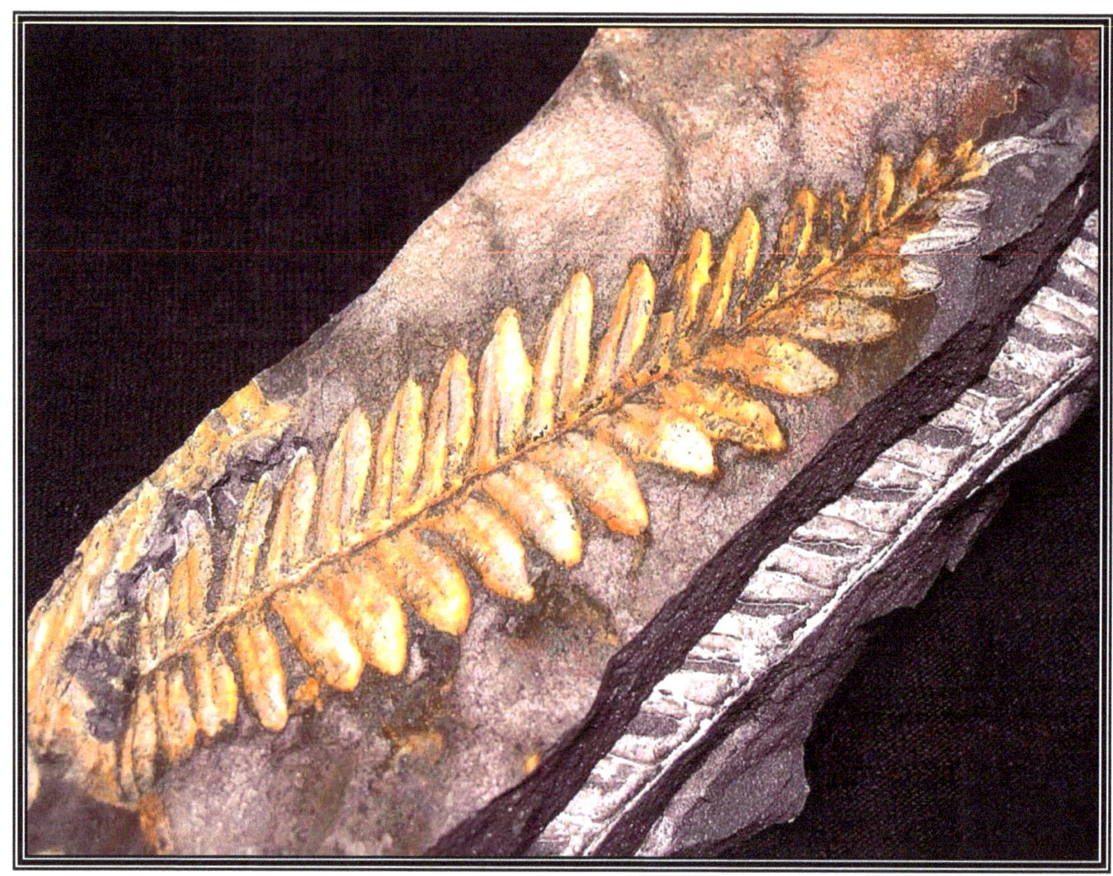

Alethopteris

Plant Fossils, Llewellyn Formation, Schuylkill County

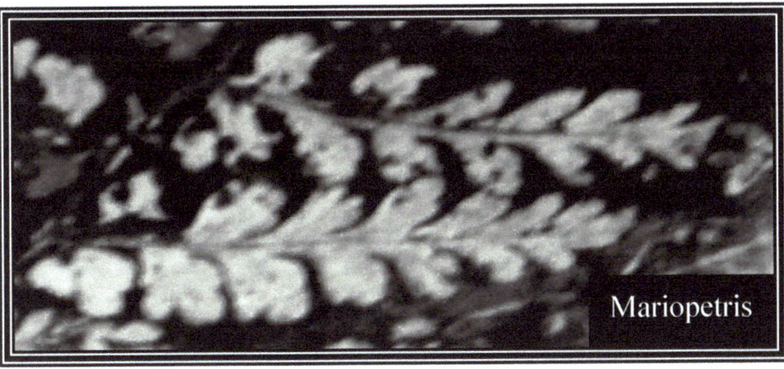

Mariopetris

Plant Fossils, Llewellyn Formation, Schuylkill County

Calamities branch

Plant Fossils, Llewellyn Formation, Schuylkill County

(74)

Fossils: Schuylkill County

Top Sphenophyllum / Bottom Rare Insect 5x

Other Fossils from Pennsylvania

Due to the scarcity of accessible sites there are other varieties of fossils collected from the Paleozoic era not commonly encountered in Pa. Not included in this documentary were many fossils of the Silurian age, (a period between Devonian and Ordovician). Pa. Silurian sites produced a few fossils, such as trilobites *Calymene*, *Liocalymene,* and a few varieties of mollusks, ostrocods (a type of swimming arthropod), and specimens of a small conical shaped creature called *Tentaculities*. A few exposures in Mississippian age rocks, which formed prior to the Pennsylvanian, produced occasional tracks of amphibians. Both Pennsylvanian and Mississippian rocks produced a variety of marine fossils in the western part of Pa.

Liocalymene Perry Co. 2009

Permian fossil remains (an age after Pennsylvanian), are found in a few places within the state yielding fossil fish teeth and plant remains. Many of these usually occur at sites less conveniently accessible from my home base. There are also fossils throughout the region of a younger age. These are usually found as rare vertebrate and plant material often in the Triassic-Jurassic boundary within the Mesozoic Era. Fine examples were collected along the Triassic-Jurassic rock boundaries, mostly in the form of fossil foot tracks from Adams, York, Northern Lancaster, Southern Berks, Montgomery, and Bucks Counties. Fossil remains from the Cenozoic era also crop up, from time to time, in Pa. Bones and teeth have been found from Ice Age mammals in peat bogs and sink holes throughout Pa.

Painted by K.Matt

Cast of a rare fossil find from the 1940's near Bowmansville, Lancaster County

Triassic-Jurassic Fossil Tracks

Pottstown Pennsylvania

Bucks County Pa.

Pottstown Pa.

Fossil Cast of a Mastodon Molar

found in a streambed near Millersville, Lancaster County. (Probably ice age)

Triassic Age Plant
found at a construction site in Bucks Co. Pa.

Index, Pa. Paleozoic Playground

section	page #
Opening pages of original book	1-5
Lower Cambrian Fossils Of the Kinzer Formation =	
Book-Lower Cambrian Explosion of life in my back yard	{has own index}
Ordovician fossils of the Martinsburg Formation=	
Book -Swatara Gap Pa. Ordovician fossil fauna	{has own index}
Fossils Bellefonte, Centre County, Pa.	39
Devonian Age Fossils, Mahantango Formation Suedburg, Pa.	40-41
Devonian Age Fossils Perry County Pa.	42-46
Devonian Age Fossils Juniata County Pa.	47-49
Devonian Age Brachiopod Slabs	50
Lower Devonian Age Shriver Formations, Centre County Pa.	51
Devonian Age Fresh Water Fossils, Clinton County Pa.	52-65
Pennsylvanian age Plant Fossils From Huntingdon County Pa.	66-67
Schuylkill County plant site	68-69
Plant Fossils Pottsville Pa. region	70-75
Triassic Age (Other Fossils From Pa.)	76-77
Ice Age Mastodon Molar	78
Bonus Book = Capturing Calvert Cliffs (81 pages)	{has own index}

Phacops, Huntingdon County

Phacops, Perry County

Produced by Kerry Matt 98% of all specimens appearing were collected in the field by Kerry Matt, Special thanks! To all land owners of fossil sites, all my mentors, The Junior Earth Science Club, and companions in the field!

CRYPTOLITHUS

THIS PAGE IS DEDICATED
IN MEMORY OF
SWATARA GAP LEBANON
CO. PA.
A FAVORITE LOCALITY
ONCE ENJOYED BY MANY
FOSSIL COLLECTORS

ISOTELUS

PHRAGMACTIS

TAENIASTER SPINOSIS

FLEXICALYMENE

Capturing Calvert Cliffs

Geologic formations in the Miocene Calvert Cliffs

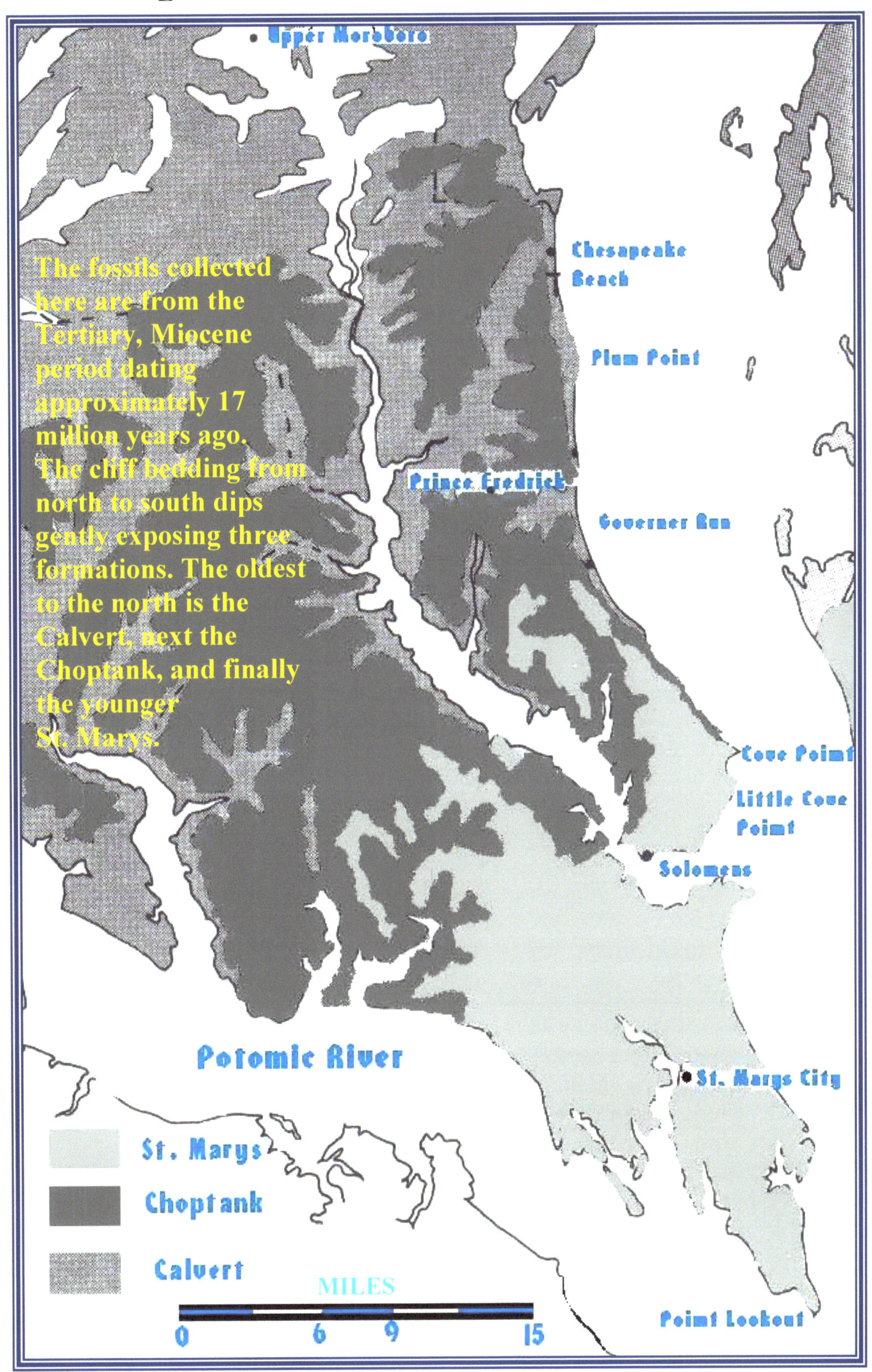

Capturing Calvert Cliffs Index

Section 1
Miocene fossil collecting at Calvert Cliffs Page 1-6
Section 2 Brownies Beach
Sunrise Page 7
Cliffs photographs Page 8-9
Carving in marl Page 10
Cliffs photographs Page 11-17
Words of caution Page 18
Tidal power and erosion Page 19
Cliff stratigraphy Page 20
Low winter tides Page 21-22
Willows Beach Page 23-24
Plaster preparation to remove bone material Page 25-26
Shell spills north of Matoaka Page 27-28
Section 3 Choptank Formation
Southern Bay Cliffs Page 29-36
Section 4 specimen material
Bivalve mollusca Page 37-43
Gastropods (snails) Page 44-47
Coral Page 48
Echinoderms Page 49
Arthropods Page 50
Fossil fish remains (teeth and bone material) Page 51-58
Salt water crocodile Page 59
Vertebrae and other bone remains Page 60-67
various unusual fossils Page 68-69
Show quality matrix specimens Page 70-72
Section 5 Matoaka
Cliffs, beaches, and fossil finds Page 73-81

This book was produced in its entirety by: Kerry Matt.
Most all specimens were collected over a period of more than 25 years by: Kerry Matt.
All photography by: Kerry Matt with the exception of a few by Rick West.
Special thanks to all fellow fossil collectors who shared the field experience with me

Names of fossils identified in this book may be outdated and changed. The Chesopectans for example may be named differently. {This book is designed to at best entertain a commonplace perspective, and is not intended as a claim of conclusive scientific data}.

Capturing Calvert Cliffs

Keeping the novice in mind I open the same way I have with my proceeding mineral and fossil books. We enter the journey from a beginner's point of view, setting the pace for new adventurers who desire to partake in field fossil collecting. Those who presently engage in this interest can enjoy comparative reminiscence of their earliest experience.

As an early 1970's member of our local college based museum's Junior Earth Science Club, the field trip experience was new to me. At the time it was not important as to what kinds of things I was going to collect in the field. It was not a significant matter as to how much, or what kinds of things I was to add to my collection. What seemed most important to me then was the fascination of just being part of a museum based team of explorers, and the excitement of participation. I was interested in learning hands-on what the natural environment had to offer in the element of surprise. I never was much of a reader, and my artistic mind in those days had a voracious appetite for independently gained knowledge

To this very day I am somewhat eccentrically the same. I peruse a small amount literary research mostly to verify and supplement knowledge of what I observe in the field and lab. To me the field is the ultimate classroom. In the Jr. Earth Science Club, the learning experience came abundantly as I look back, yet at the time it all seemed mostly like fun attuned to admiration of the outdoors and the natural environment. Field trips to various places would come and go. I developed shelves full of specimen material including rocks, minerals, and fossils to display as my own mini museum collection. In the coarse of events with the club, I recall chatter among individuals of great fossil finds form Plum Point Maryland. Talk of expeditions made along Calvert Cliffs western shore of the Chesapeake Bay. I heard about porpoise skulls observed hanging in cliff walls, and large fossil teeth from sharks measuring several inches worth over $50.00 (a fair chunk of change for a kid in the 1970's).

The time finally arrived for a plan to take a bus trip to collect marine fossils in the Choptank Formation of Calvert Cliffs. We set our date on our calendars. For a few dollars and a permission slip from our parents it was a sure go for myself, in company of a neighbor friend. This time our destiny was in a new area, a large electric generating power plant facility in construction along the waters of the Lower Chesapeake Bay. Apparently Plum Point known as a fossil hunters paradise for vertebrae and teeth was no longer accessible. I remember a gray cloudy day, climbing aboard a yellow school bus, heading to a destination I'd ever been before as one who was a reserved quiet type among a group of chatty adventurous kids

After several hours finally arriving at our destination we turned down a road to meet our guide, then proceeded to exit the bus heading down a path through a wooded area leading us to the shoreline of the bay. I was captivated at the sight of the steep clay cliffs. Gray-clouded sky's persisted appearing as though it might rain at any moment. I can still recall voices reminding of those big Carcharodon Megladon teeth and potential for big bucks. Enthusiasts scurried down the beach disappearing as if knowing there was a reason to be the first out there. Being totally new at this I didn't really go too far or speedily as the rest of the bunch. I was amused with large scallop-like pecten shells hanging on the cliff walls as I grabbed and wrapping bunches of them. I puzzled over how modern looking the preservation was. Some were decorated with oversized barnacles. The perfectly articulated clay filled bi-valves seemed to be abundant. These are spectacular! I thought, seeing these scalloped hand size mollusks for the first time. Are they actual fossils? Or am I finding something left over from yesterdays wash?

Later I learned they were true fossils of the Miocene period dating around 17 million years. It was hard to believe these were fossils. They appeared pristine compared to the Paleozoic rocky steinkerns we were used to collecting back home.

Capturing Calvert Cliffs

At the time I had no prior experience hunting fossil sharks teeth, but I now under stand reasoning behind many in our group hastily making their way down the beach. Experienced collectors were accustomed to spotting the wash in the surf and beach. They were looking for accumulative fossil remains in the agitation zone, which is the most popular method. I was exploring the wall surfaces, and was proud when finally finding my first half-inch, black shark tooth. I had no idea which kind of a shark it came from until later years. It was from a snaggletooth shark *Hemipristis Serra*. Of course time flies and it seemed the duration of the collecting was short. I packed my collectables into my carrying bag as we were rounded up for our hike back to the bus. I barely remember the ride home. Most of the memory (some 35 years ago) is as misty as the weather finally arriving back at the museum.

Though many years have passed since this inspirational field trip the tooth I found survived, bearing trace remnants of purple and green mounting clay along its root. The *Hemipristis Serra,* among many to be found later, still remains in my collection today. The pectens, and whatever else I collected on that trip have long since crumbled.

As years with the Jr. Earth Science Club passed and ensuing high school years at least half a decade further on my interest in earth science endeavors revived. A time had finally arrived when I felt the desire to get back to exploring the Miocene fauna once again. My primary interest and knowledge in fossils and minerals of Pennsylvania were advancing. Now I was at an age equipped with my own transportation and exploring opportunities were wide-open. On a path to becoming a dedicated representative of Pa. faunal, Miocene age vertebrate material was geographically over the boundary line. Nevertheless collecting at the bay became more of a recreational luxury incorporated within my primary interest. Now comes a different age for me exploring the Calvert Cliffs in the mid 1980's

Beach at Calvert Cliffs State Park

Capturing Calvert Cliffs

Enthusiasm over fossil collecting potential of Maryland's western shore inspired my perseverance for locality data. Inquisition and research became a necessity before I was to get my fossil fix in the vertebrate class. The element of tress-pass became a booming, or better yet, bitter issue in the picture as the 1980's progressed. The good old days of freelance access to sites were rapidly crumbling. A few booklets and good words from experienced fossil collectors set the pace for my future quarry. It was time to build a collection representing what a three-hour drive could produce in fossil fauna that Pa. soils could not.

First and most interesting was finding out about Calvert Cliffs State Park. My initial thought was hurray for who ever was so considerate of humanity in light of our freedom to roam what mother nature gave to us all. To preserve such a place for the public to enjoy was important. I was, and still am offended that shoreline of any body of water so great as our Chesapeake Bay could become private property. The main issue though comes down to access. I later learned if one can simply get to the shoreline by what ever means, less treading upon private thoroughfare, it is legal to be on ground up to the high tide mark. In many cases with a boat the Bay Shore is your own unless posted off limits by the local government. Thanks to Calvert Cliffs State Park I found to my delight, a place guaranteed to fit my budget for a bay fossil trip.

The down side (if I dare call it that) is a hike of at least one and three quarter miles (three plus miles round trip) was required before any fossil hunting was to take place. This cuts into collecting time, however the exercise and natural scenery along the trail was well worth it. In the early days of my visit sharks teeth seemed to be fairly abundant. It was easy to find several species of teeth up to an inch or more, crab claws, and stingray teeth were plentiful. Another disappointment, exploring the cliffs was not allowed for safety reasons. Here again the coming of a new age where liability is a major issue. This made things difficult for accumulating much desired mollusk shell material. My needs in collecting were pumped up well beyond being content from a novice perspective.
I wanted to collect complete specimens or none at all. I enjoyed several ventures to CCSP and was once allowed to access another part of a beach nearer the cliffs after voicing my deep interest to those minding the premises. I found a large lower *Hemipristis* tooth and a good plate from an eagle ray to highlight my collection. I encourage nature lovers and enthusiast to use and enjoy the beauty of this park. With more research I needed to find good mollusk material to build a collection. Meeting other avid enthusiasts at the park was inspiring. New ideas for other collecting sites evolved from there. It was time to head north to Matoaka where for a fee you could access cliff areas north of Calvert Beach.

Looking north from Calvert Cliffs State Park Beach photo by: Rick West

Capturing Calvert Cliffs

Matoaka soon became a favorite locality where cliffs north of Calvert Beach supported grounds reported to have been a former Girl Scout camp. It is now a privately owned resort accessible to fossil collectors. Owners will rent cabins, allow primitive camping, or a day on the beach for a very reasonable fee. It is always made clear to visitors that digging in the cliffs is strictly forbidden. My experience there clearly demonstrates cliff digging is not necessary at all for good collecting. Loose blocks often separate from the cliffs and are accessible. Collecting is allowed from this material. The idea is to protect and discourage erosion to the precious land.

My timing was good during my first visits. Large blocks had spilled away from the cliffs presenting layers of bedding from rich stratigraphic zones. The zones yielded a vast assemblage of mollusk material harboring vertebrate fossils as well. Most enthusiasts practice a method producing many fossil teeth and other small isolated specimens. They simply stroll along the shoreline with a sharp eye and a quick hand. Some will do this with a long handled sieve or a metal scoop designed just for this purpose. I found many teeth and small vertebrae this way. I custom designed a two-foot wide, quarter-inch mesh device for this style of collecting. I also heard of putting clear plastic on the bottom of a five gallon bucket to peer into cloudy waters for spotting teeth on the bottom of the agitation zone. Another trick was getting on the beach at the crack of dawn being the first to spot whatever the tide might wash up. The best of all was using a facemask and diving for teeth. I once saw an individual doing this sporting an onion bag filled with two to three inch teeth. I tried this method but had a hard time getting the hang of it. I found a good atlas vertebrae covered with modern barnacles that needed to be scraped off.

My favorite, and best method is collecting from loose block. I find there is no better way to display fossils than those appearing in their original matrix. Aesthetics are important to me. I select outstanding specimens with multiple species potential, taking them home and working on them in the lab. It was common to find teeth, vertebrae, and other surprises mixed in with the mollusk shell material while dissecting clay boulders. I usually enjoy spending time working in a given area versus beach patrolling.

Capturing Calvert Cliffs

The Chesapeake Bay environment provided much pleasure partaking in fossil hunting. Long warm summer days in the sun with one of the best natural views on the east coast created many great memories. I continued yearly visits to this locality. A stroll exploring the beach, the white sands, high cliffs, and greenery of banks covered in alluvium induced tranquility under a big picturesque sky. I often referred to this area as the Little Ivory Coast of the USA

Time changes things and the learning continues: One thing I found pleasing was the color contrast in the material I collected. The shells had an interesting beige-pink color contrasting a dark blue-green matrix. The clastic nature of the cliff is composed of clay or marl. I discovered this fades rapidly after the material dries out. The good news, if this material is barely ever handled and stored in the proper environment it seems to hold up indefinitely. Many matrix specimens shown in this book are from my earliest collecting times. The down side considering aesthetics, blue-greens become gray and pinks fade to dead white. Another important observation is how much the cliffs and beach can change. All the great luck I experienced on my first visit to Matoaka was taken for granted. A few summers past and I found erosion had long since taken all of the boulders away. I remember a long period where it was tough to even find a whole pecten shell let alone much of anything else. Big perfect articulated bi-valve scallop shells that could be collected by the bushel basket became scarce. After major storms the beach can exhibit a great deal of change. I remember in the 1980's the northern part of the cliff area at Matoaka was precarious and narrow. The tide washed right against the cliffs. It was slippery and there were cobbles, but little beach. In the 1990's a major hurricane left the same area where today a sandy beach now exists as wide as a soccer field.

Other seasons came again where there were great shell blocks to pick through. The lesson is to stay on top of this situation and return as often as you can while the material lasts. Not only does erosion claim the fossils. I have watched the interest in this Paleontological hobby grow into enormous proportions. Collectors will devour the fallen block as quickly as Mother Nature herself.

The pages of this book exhibit fine specimens of material taken from block along this shoreline. There are good examples of common mollusca and a few less common. I exhibit sharks teeth preserved in matrix and vertebrae as well. Great care was taken to preserve specimens in the lab showing them as they naturally appear. One must learn lessons from mistakes in capturing bone material. I experienced times underestimating how deep to cut into a block to remove a fossil and came out short changed. It is important to take care removing everything intact if possible.

Preparing material at home is important and must be done with the right tools.

Prior understanding of vertebrate taxonomy can be of great help. I learned this the hard way with a few whale vertebrae that could have been collected complete. They can also be extremely fragile. One of my most interesting finds appears to be a partial skull from a large mammal.

I found this by carefully examining block. Most times this kind of discovery may only appear as a fragment (a few percent) of a bone mass visible on a surface. You never know if you may have a large complete skull or whatever. A hasty excavation can ruin a prize specimen. Sometimes it is worth the extra time and manpower to carry out extra weight for a lab preparation. Hind site cannot restore lost parts of a prize find.

A favorite gastropod (snail shell) to collect, often reasonably abundant at Matoaka, is the Maryland State Fossil *Euphoria Quadracostata.* I enjoyed collecting many of these in matrix when ever I could. One of my most memorable days was during a weekend camping trip. Equipped with a boat, we had little success during beach exploration along the southern coast. The evening of our return after finding nothing I decided to spend the last half hour of the day searching the tide. I spotted a strange crescent bone in shallow water among cobbles along the northern narrows. I picked it up and ironically it turned out to be a four-inch Carcharodon tooth. Within the same half hour I picked up a two-inch crocodile tooth. It felt special to sport these on my walk up to the cabin for the evening's retirement, jumbo shrimp, and a few cold brewskey's.

Capturing Calvert Cliffs **Brownies Beach**

Visits to the cliffs at Matoaka were productive in compiling a noteworthy collection. Especially while considering my distance and inability to visit as frequently as local enthusiasts from the lower bay region. There were always a few new unanticipated additions for the collection along with repetitive finds. As years passed I began to notice a decline in the amount of beach collectables available. I could see how the enthusiasm over fossil hunting was becoming a rapidly expanding hobby interest for many people. Before I had a chance or reason to become discouraged came another surprise.

A friend from local rock and mineral clubs informed me about what at the time was generally a well-kept secret. It was a place known as Brownies Beach. He invited me, introducing me to this collecting site at a time I would never have guessed. We left in the pre-dawn hours equipped with our warmest winter clothing and hip boots for a winter expedition. It was New Years Day and I can't remember the exact year, but I think it was in the 1990's. The Idea was to get out on the beach at the crack of dawn being the first to spot any new fossil remains that might have washed ashore.

When we arrived there were several inches of snow still around but as the sun rose on the bay I was amazed to see the beach was clear. The sun shining off of the water against the cliffs had a warming effect and collecting was comfortable. The entire experience was exhilarating to walk the path to the beach seeing this place for the first time. I was delighted by the fact this beach was the northern part of Calvert Cliffs, at least a half hour closer to home. The numerous pickings seemed easy at this site and I soon discovered the abundance of shark teeth matched Matoaka a decade in its past. I was able to collect a few whale vertebrae and rib bones frozen in a section of block. My partner found a similar specimen of good quality surrounded by matrix.

View south of Chesapeake Beach

Starting a new day of searching at Brownies Beach

Photo by Rick West

Capturing Calvert Cliffs **Brownies Beach**

Today Brownies Beach with paved parking area and access trail is now a park dedicated to fossil hunting and public beach activities. There were many good finds over the years since my first visit. Winter is the best time for fossil hunting with much lower tides and easier access. I have been there during warmer months where the water was up against the cliffs and barely accessible. This area can be risky and one must be on their toes at all times as to where you are standing along the cliffs. Blocks of enormous size can fall at any time.

This locality occurs in the Calvert Formation (oldest, Miocene age bedding). Matoaka and Calvert Cliffs State Park graduate into the Choptank formation. I liked Brownies Beach the best for teeth and vertebrate material. Mollusk shells at this locality are usually found in poorly preserved condition with the exception of oyster shell layers near the base of the cliffs. The beach area is excellent for sifting and safe for youngsters.
The abundance of many types of smaller teeth is pleasing and sifting will often produce porpoise teeth. Digging through loose block and spill zones can be quite productive.
I have witnessed individual's producing rather nice Megladon teeth this way.

This area has also become very popular with fossil collectors over the years and seems to have shown a decline in numbers of larger specimens. Nevertheless a few usually always do turn up with each visit. I discovered in recent years, summer months have become challenging to fossil collectors unless you arrive early. I have seen this beach packed full of sunbathers, barely able to find a place to park on a mid day visit.

View north towards Chesapeake Beach

Capturing Calvert Cliffs **Brownies Beach**

View south

Carving in clay boulders along cliffs south of (Brownies) Chesapeake Beach

 Brownies Beach

View north

Capturing Calvert Cliffs **Brownies Beach**

View north showing cliff erosion

View south showing large spills and slumps

Capturing Calvert Cliffs **Brownies Beach**

View north to Chesapeake Beach

Capturing Calvert Cliffs **Brownies Beach**

Capturing Calvert Cliffs **Brownies Beach**

Capturing Calvert Cliffs **Brownies Beach**

Capturing Calvert Cliffs stratigraphy in the Calvert Formation, Brownies Beach

Capturing Calvert Cliffs

This is something important to watch out for. You don't want to hang around here

Always be cautious. Beware of conditions above. You don't ever want to end up looking like this

Here is an example of tidal power and erosion from a wooden beam against the cliff.
Digging in cliffs is prohibited: one small hole can cause a great amount of erosion

Capturing Calvert Cliffs Stratigraphy: upper and lower (oyster shell bedding) **Brownies Beach**

Fossil beds, species Ostrea

Capturing Calvert Cliffs   Brownies Beach

View south low winter tide

View north low winter tide

 Brownies Beach low winter tide (view north)

Capturing Calvert Cliffs

Another collecting opportunity was being a guest at Willows Beach. This area is privately owned but I was permitted to visit with a club. There was an area where I was able to collect numerous seven-gill (*Notorhynchus*) and cow (*Hexanchus*) shark teeth. I used a long handle screen in the shore-wash off of the point.

Expeditions I've been on in recent years were often by boat. This too seems to be a popular method used by the serious collector. The beaches seem well picked over and I try to imagine what it must have been like 100 years ago. Were the beaches littered with teeth and bones? Or is it just speculation? Access by boat has allowed me to make more fine additions to the collection form Choptank areas in lower parts of the bay.
I finally had the chance to see Plum Point and other areas as well. I am now more familiar with most all of the cliff areas.

One part of Miocene Period fossil hunting I have not yet experienced was to work in the St Mary's formation of the lower bay. I heard reports of fossil starfish finds in limonite concretions, fossil crabs, and many types of gastropods. This might make a good future project to expand upon.

With all of my experiences I think I have plenty to share with the many photographs used to document most all aspects of Miocene age marine fossil collecting. However I have barely exhibited the tip of the iceberg. There are several hundred species described from Calvert Cliffs. I exhibit many common specimens that can be accumulated over time by the average fossil collector. I plan to continue visits to sites and hope I may find a complete fossil skull in a block one day. I have observed skull and bone material in the cliff face and have practiced the ethics of reporting these to the Maryland Calvert Marine Museum. I hope all readers who decide to engage in fossil collecting will do the same helping to share important information with the scientific community.

View north from point at Willows Beach

CapturingCalvertCliffs Willows Beach

Willows Beach view south

Willows Beach view north from point

Plaster casting by Calvert Marine Museum on suspect find of whale remains.

Leave this type of digging to the experts. **Report suspect finds for proper excavation.**

Bone and possible skull casts from a slab found broken away from the cliff base. Expert paleontologists are alerted for a possible good find.

Capturing Calvert Cliffs north of Matoaka: Scientist Cliffs

View south toward Matoaka and Calvert Beach

Block full of shell material around Scientist Cliffs area

View South toward Matoaka and Calvert Beach

Southern cliff areas in the Choptank Formation

Southern Cliffs of the Choptank Formation

2" Carcharodon Megalodon tooth untouched photographed as found in place on beach wash

Boulder conglomerate full of chesapecten shells

Cliffs South of Plum Point

Calvert Cliffs Choptank Formation

4" Carcharodon Megalodon Tooth as discovered by the captains wife on a boat expedition in the southern cliff area (Choptank Formation).

This was a first find just as the boat landed. The rest of us hastily scattered left and right missing this one sitting on the beach directly in front of the boat. This made one of those great collecting memories

Calvert Cliffs Choptank Formation

A bone is spotted immerging from a clay mass in the floor appearing to be whale vertebrae

Calvert Cliffs Choptank Formation

A gritty job on a hot humid day

Partial excavation in progress

Tough gritty mission accomplished

Partially exposed Ecphora quadricostata

Boulders composed of shelly conglomerate

Whale vertebrae specimen from lower bay cliff area

Specimen after preparing in the lab

Opposing side of prepared specimen (page 35)

Bi-Valve Mollusca: chesapecten (view of both sides)

Bi-Valve Mollusca: chesapecten in conglomerate Lower Bay Choptank Formation

PECTEN An uncommon species pictured above

A possible different species of pecten showing subtle surface variation

Scalloped Bivalve Mollusca:

CARDIUM

ARCA

Big Clams

PANOPEA

PANOPEA

Common Mollusk Shells From Calvert Cliffs

GLYCYMERIS

EUCRASSATELLA

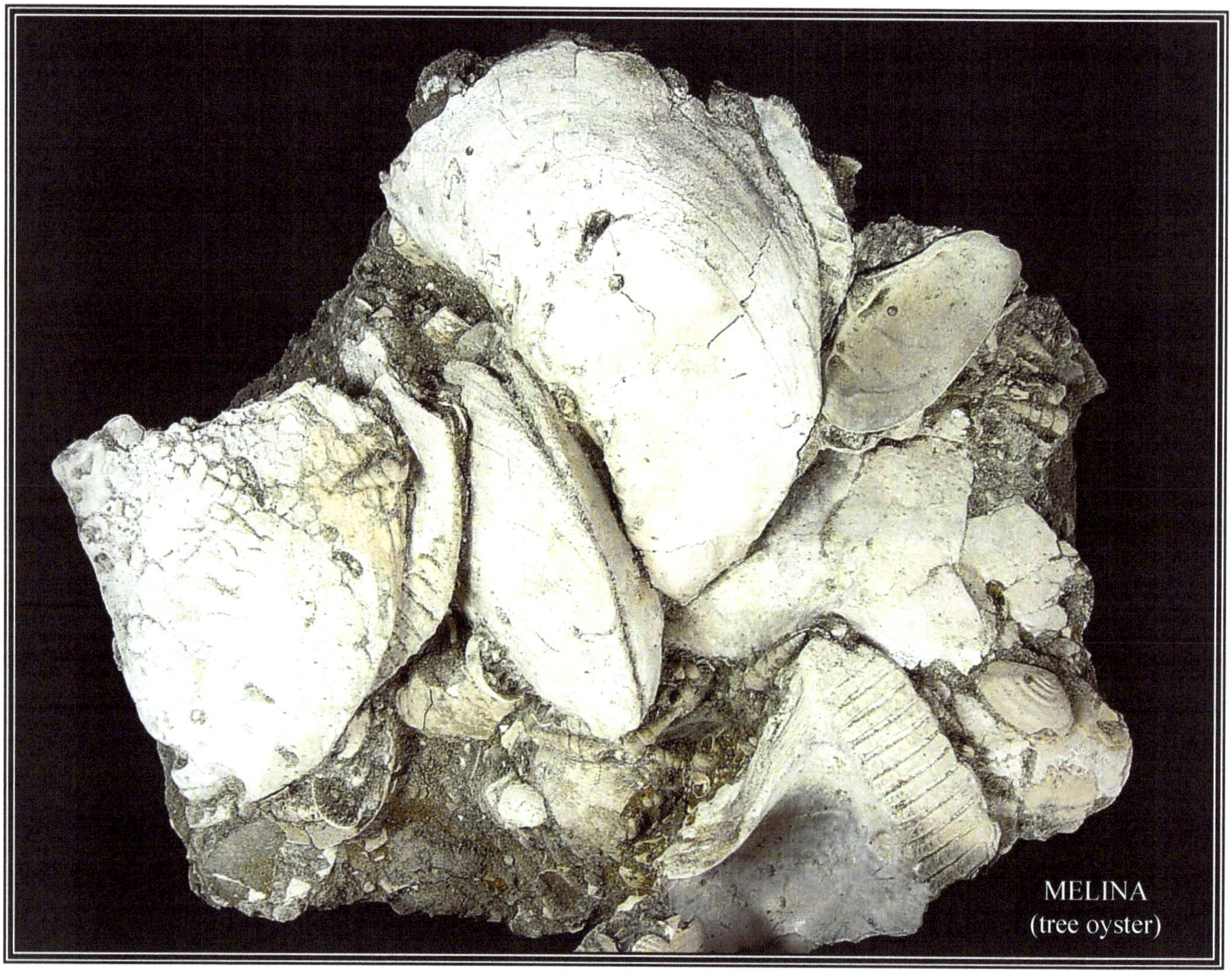

MELINA
(tree oyster)

OSTREA (Oysters)

(43)

Gastropods (snails)

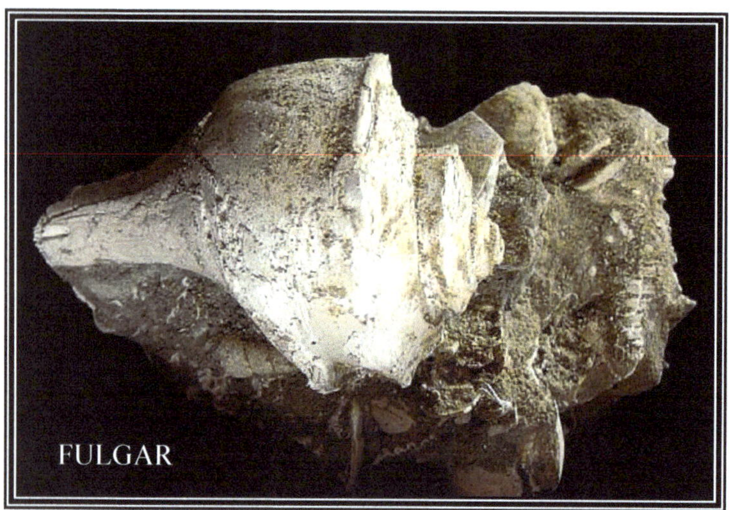
FULGAR

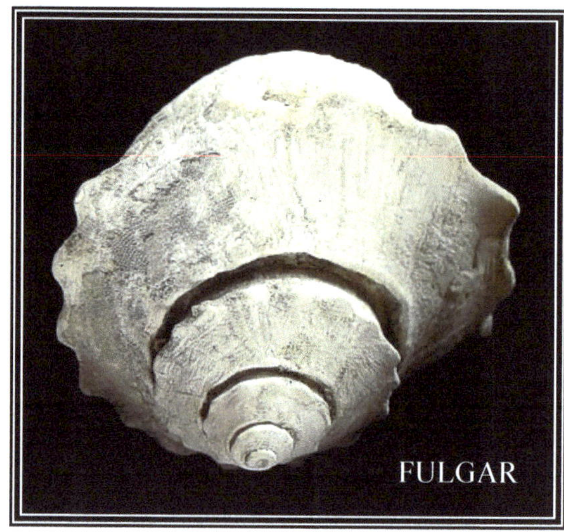
FULGAR

TURRITELLA

SIPHONALIA

TURRITELLA 2 different species

Gastropods (snails)

SCAPHELLA

SCAPHELLA

POLYNICES with turritella and shark vertebrae

TURRITELLA

POLYNICES

Maryland State Fossil: {Gastropod} Ecphora Quadricostata

Maryland State Fossil: {Gastropod} Ecphora Quadricostata

ASTERHILIA (Coral)

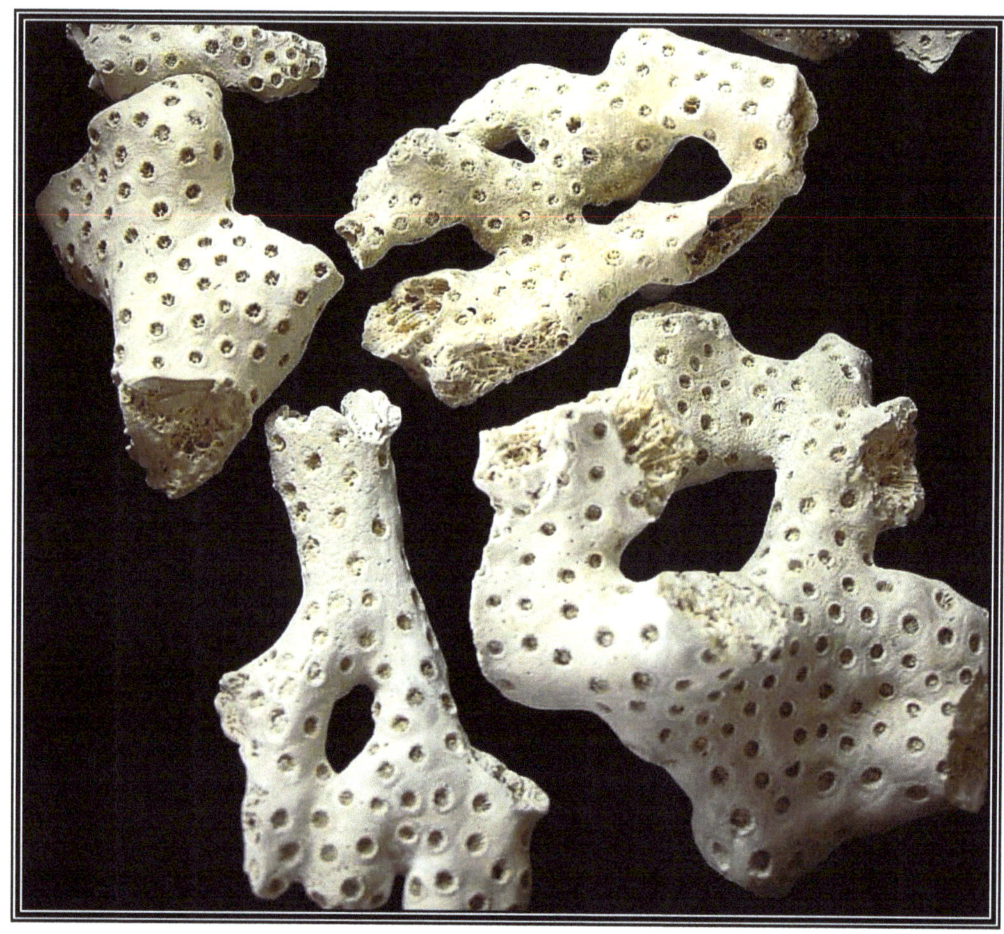

Echinodermata SCUTELLA (sand dollar)

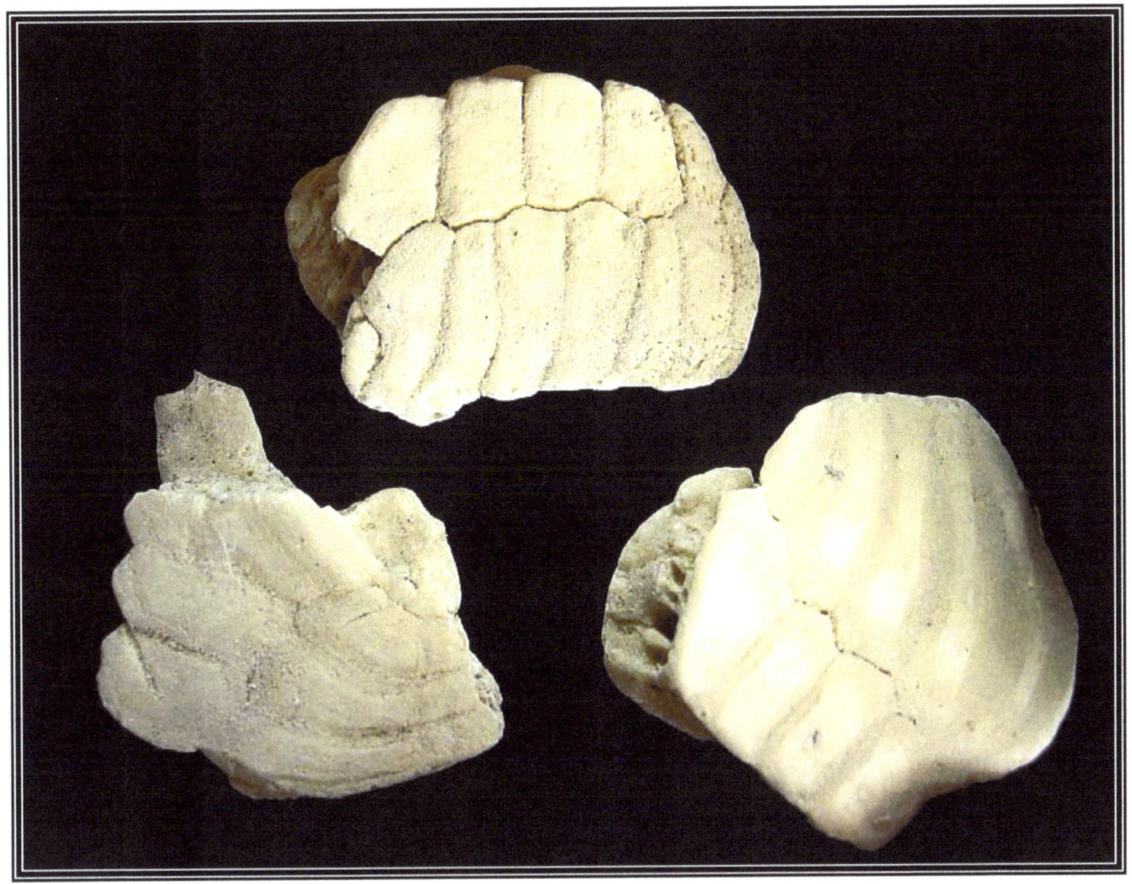

Arthropods: barnacles and crabs

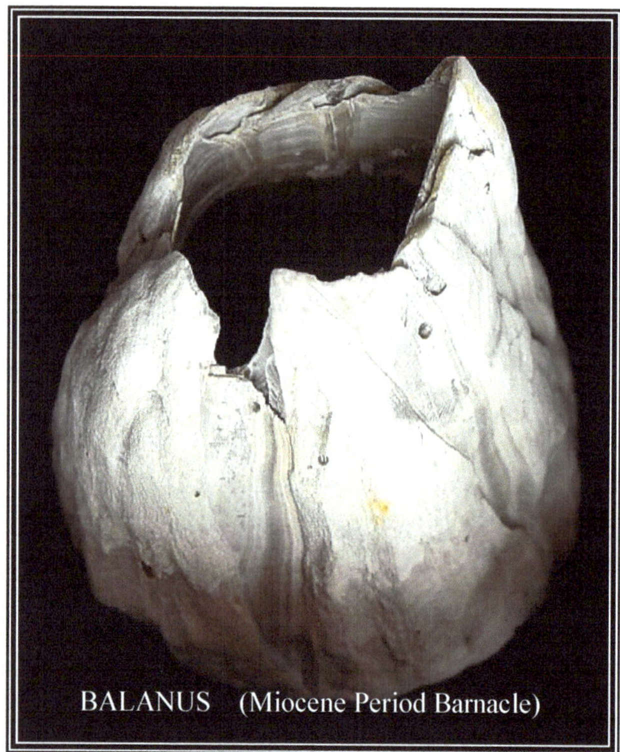

BALANUS (Miocene Period Barnacle)

The hard claw parts are usually all that remains of fossil stone crabs from the Miocene of Calvert Cliffs. (Common in the Choptank)

BALANUS

Fish vertebrae / shark centrum

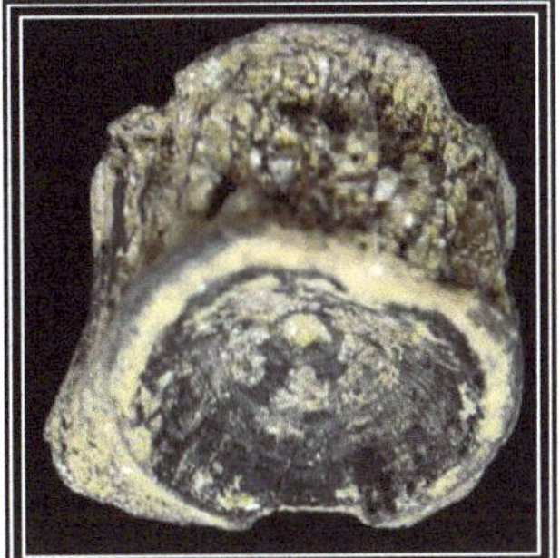

Miocene shark teeth from Calvert Cliffs Maryland

GALEOCERDO Tiger Shark

GALEOCERDO Hammerhead Shark

Miocene shark teeth from Calvert Cliffs Maryland

CARCHARINUS Sandbar Shark

ISURUS Mako Shark

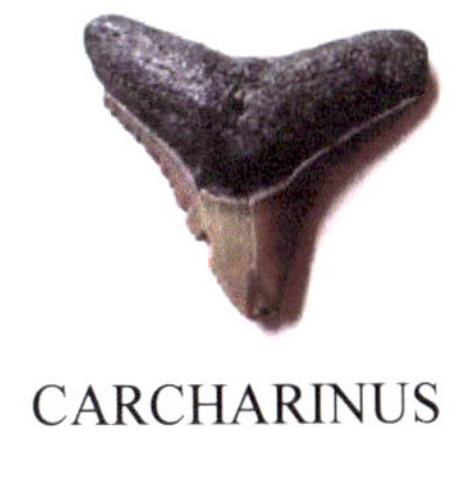

CARCHARINUS

HEXANCHUS
Cow Shark

OTODUS

CARCHARODON juvenile

ALOPIES
Thresher Shark

CARCHARINUS

needs identification

Miocene shark teeth from Calvert Cliffs Maryland

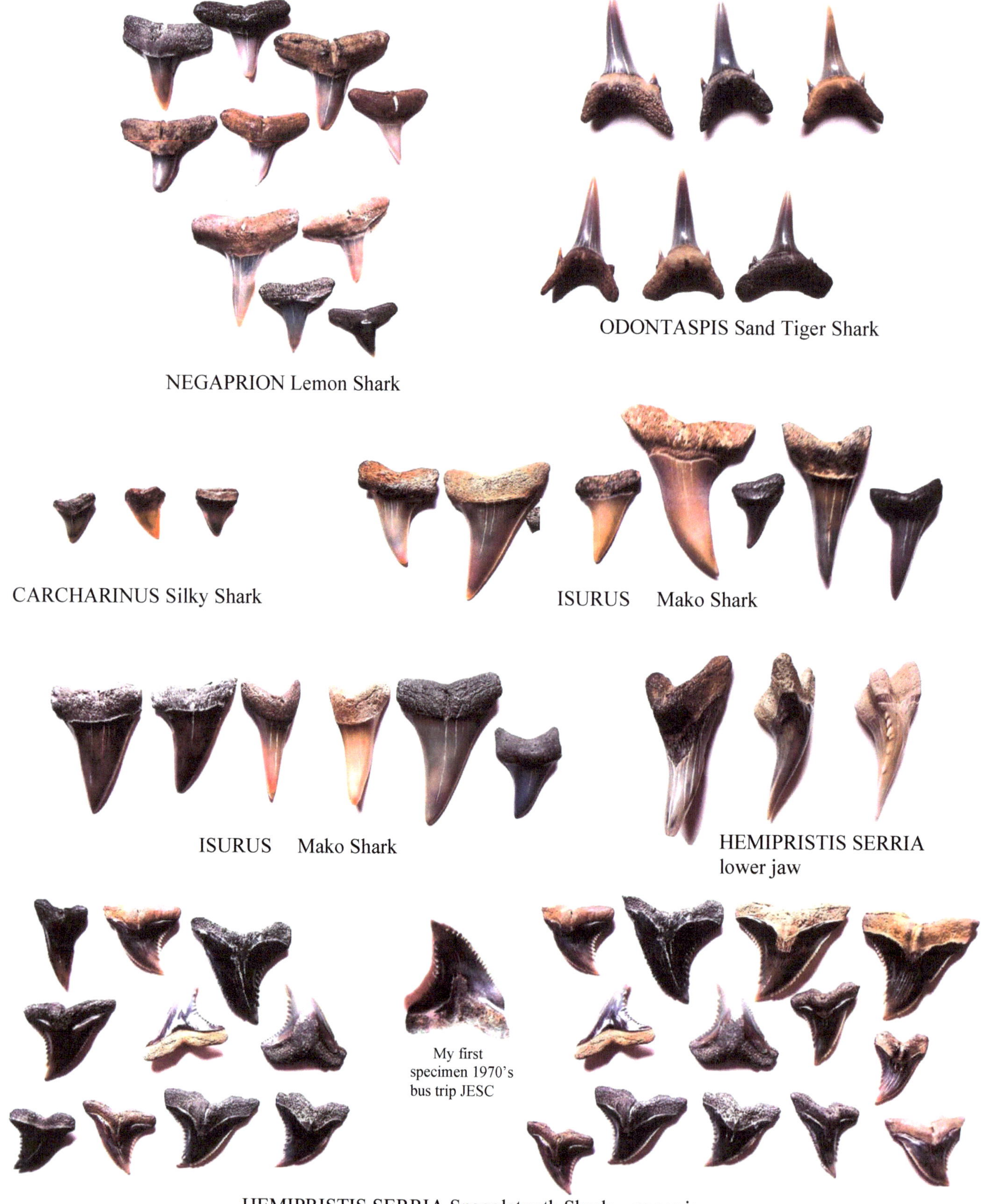

NEGAPRION Lemon Shark

ODONTASPIS Sand Tiger Shark

CARCHARINUS Silky Shark

ISURUS Mako Shark

ISURUS Mako Shark

HEMIPRISTIS SERRIA lower jaw

My first specimen 1970's bus trip JESC

HEMIPRISTIS SERRIA Snaggletooth Shark upper jaw

NOTORHYNCHUS Seven-Gill shark teeth

Blown up approximately 4x

Prize specimens of shark teeth from Calvert Cliffs

CARCHARODON
MEGALODON
Matoaka Beach

Great White

ISURUS (Mako)

ISURUS (Mako)

Lower Hemipristis Serra

ISURUS (Mako)

HEMIPRESTIS SERRA upper

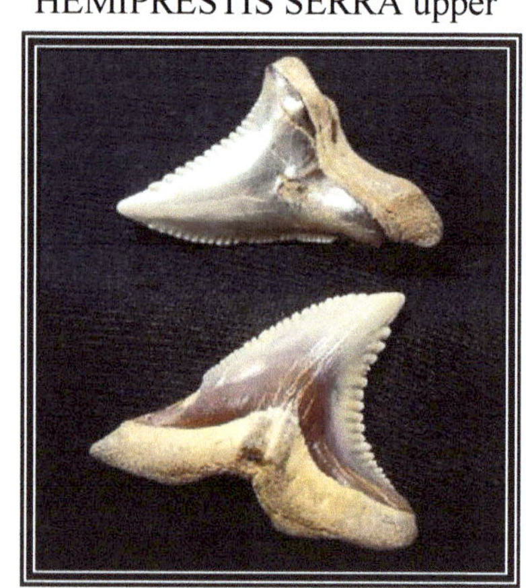

Various sharks teeth in framed mount

Stingray teeth, dental pavement, and spines

MYLIOBATIS

Salt water crocodile teeth, dermal plates, and tortoise shell

Tortoise shell fragments

Whale vertebra punctured by crocodile teeth.
Found in a northern stretch of beach at Matoaka where all Crocodile teeth appearing here were collected.

Thecachampsa
Salt water crocodile teeth

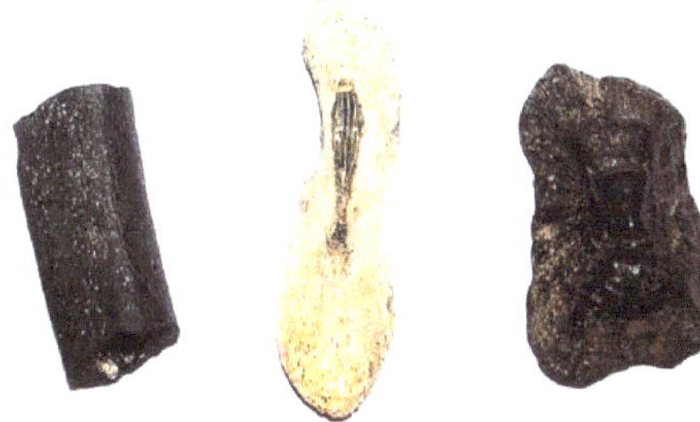

Reptilian dermal tubercle
(skin surface plates) and bone

Bones and vertebra of large aquatic mammals

(60)

Bones and vertebra of large aquatic mammals

Miocene Period Whale Vertebrae Calvert Cliffs Maryland

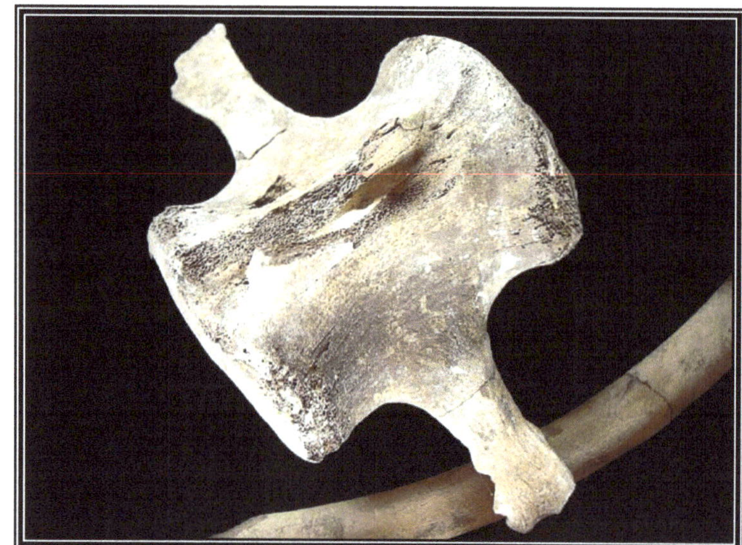

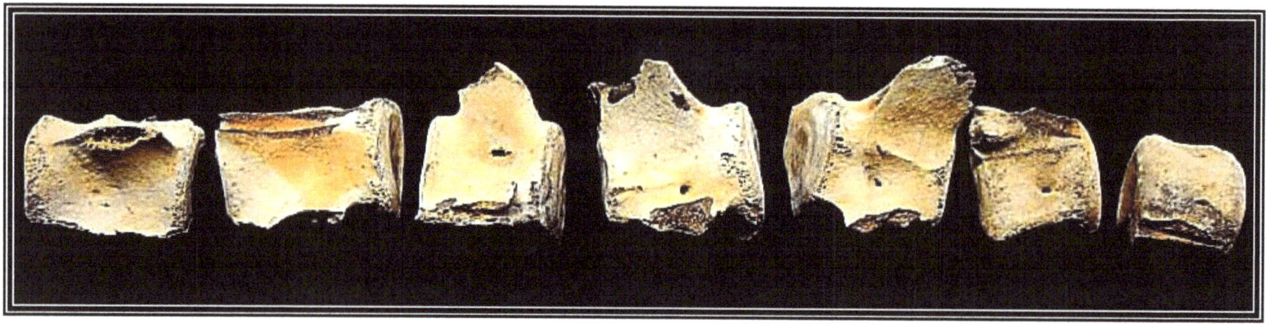

Bones of large aquatic mammals: whale vertebra

Matrix specimens: Lower Bay, Choptank Formation

Various types of vertebrae

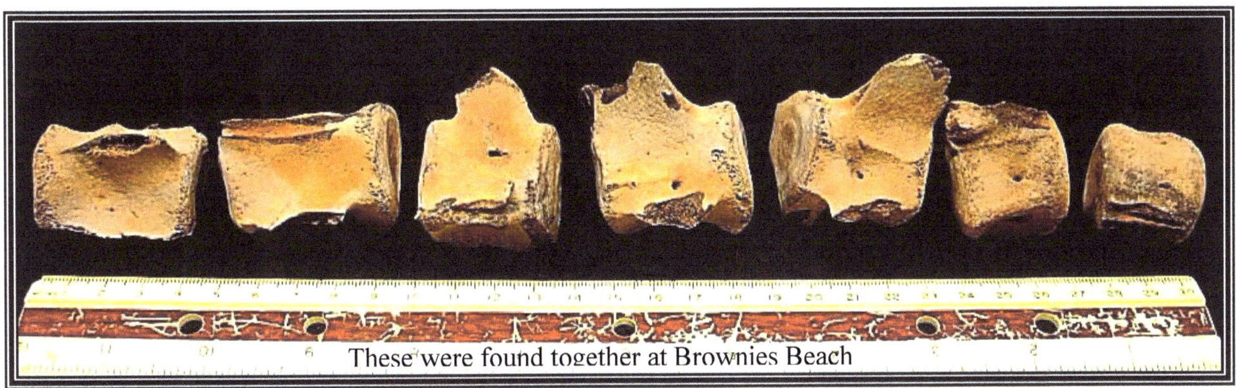

These were found together at Brownies Beach

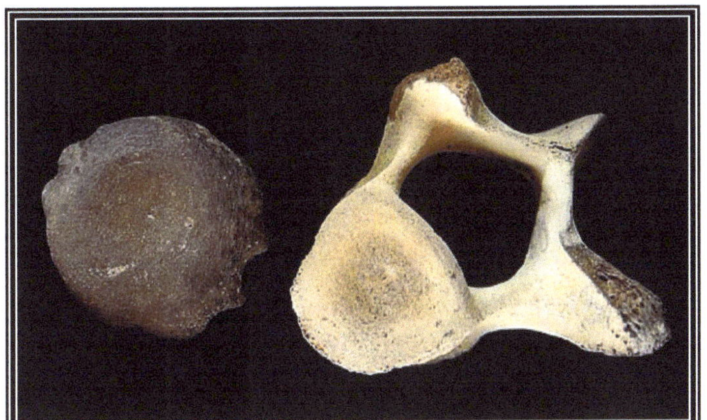

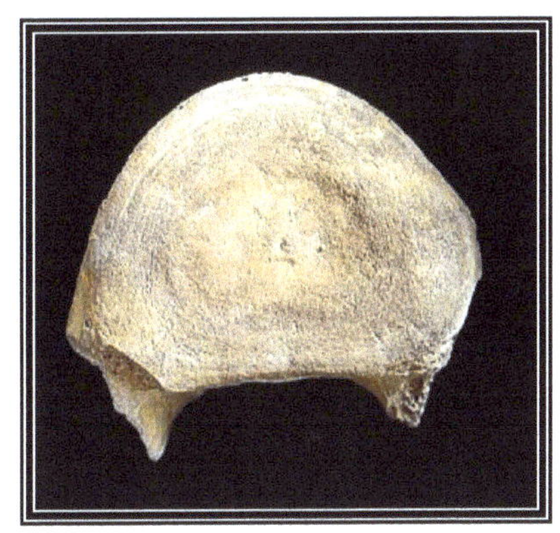

Miocene age vertebrate material, Calvert Cliffs Maryland

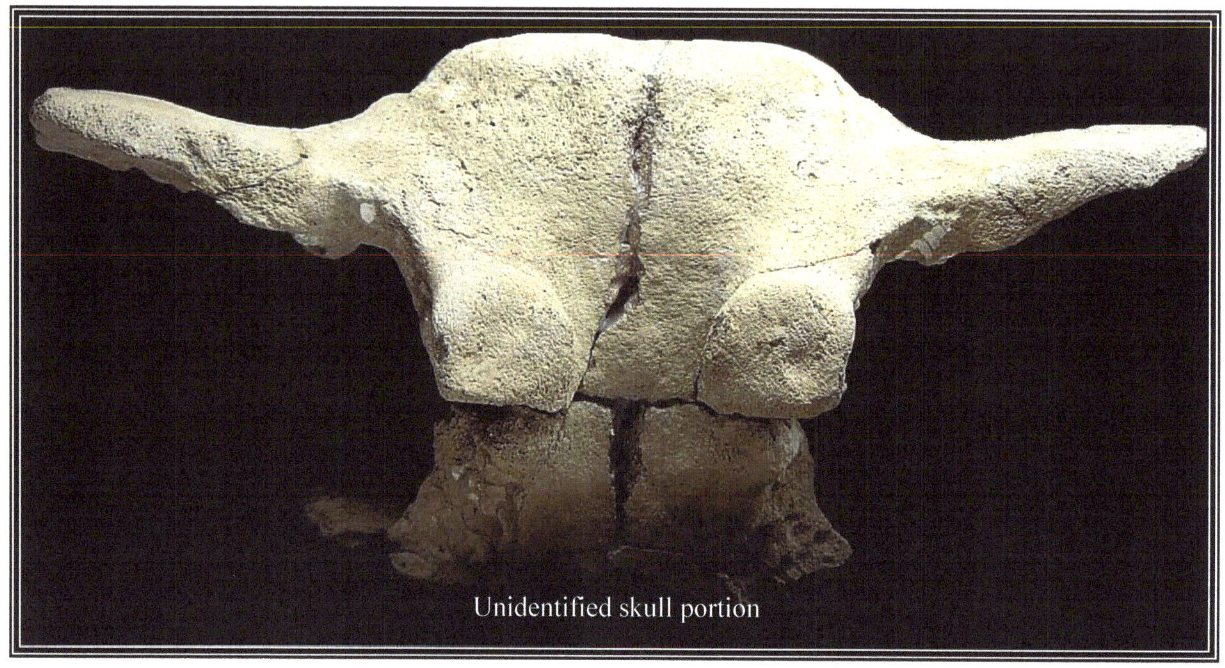
Unidentified skull portion

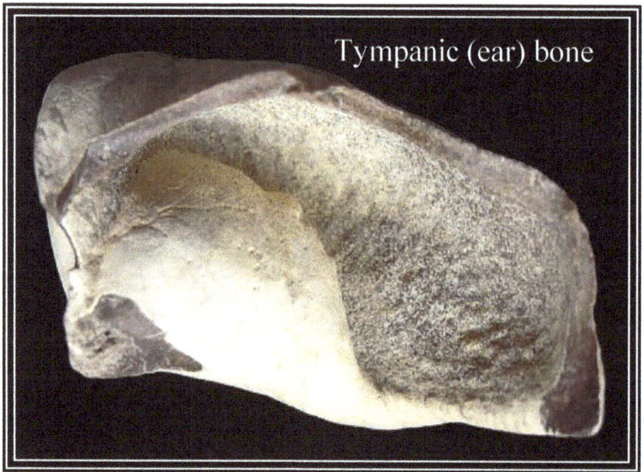
Tympanic (ear) bone

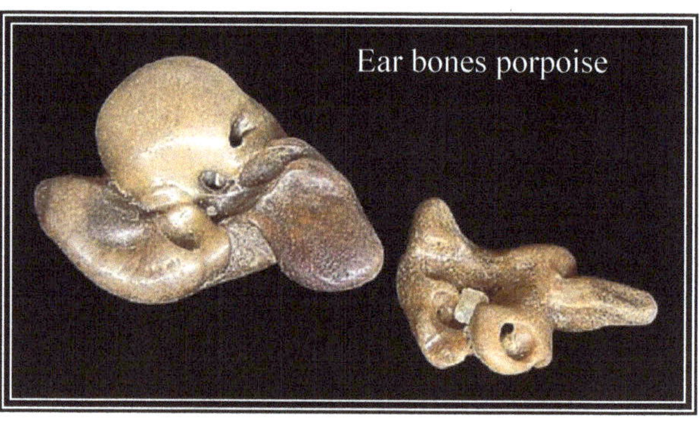
Ear bones porpoise

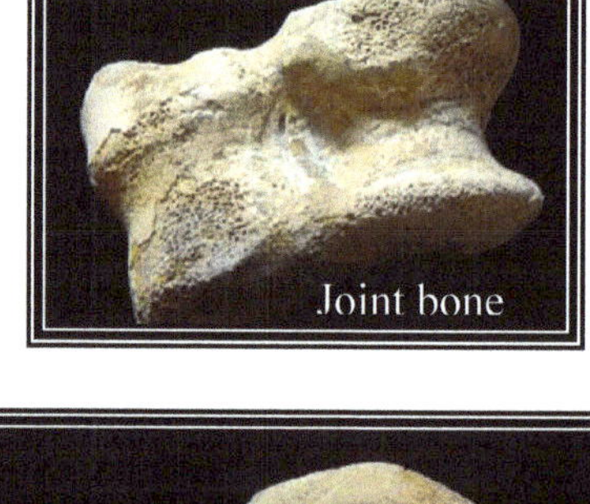
Joint bone

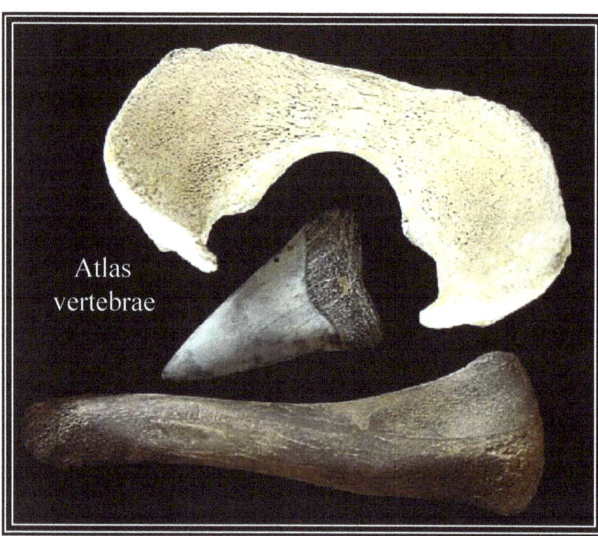
Atlas vertebrae

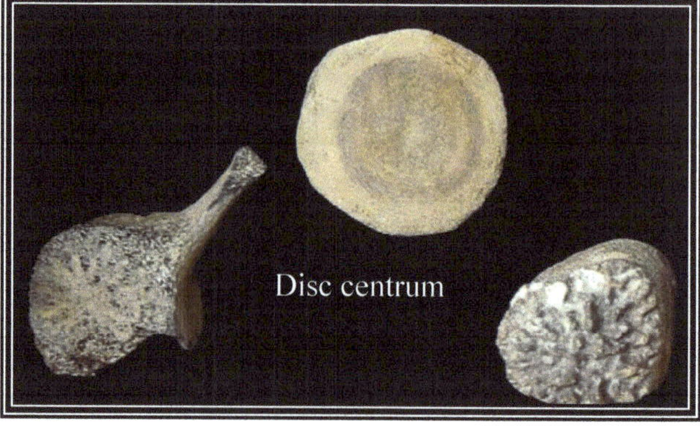
Disc centrum

Porpoise skull in cliff showing jaw parts in place: Brownies Beach

Porpoise Teeth Rhabdosteus and Delphinodon

Unusual specimens: Miocene, Calvert Cliffs Maryland

Fossil sponge like animal 5x

Lignitized nut or seedpod 5x

Unusual assortment of tooth and bone material

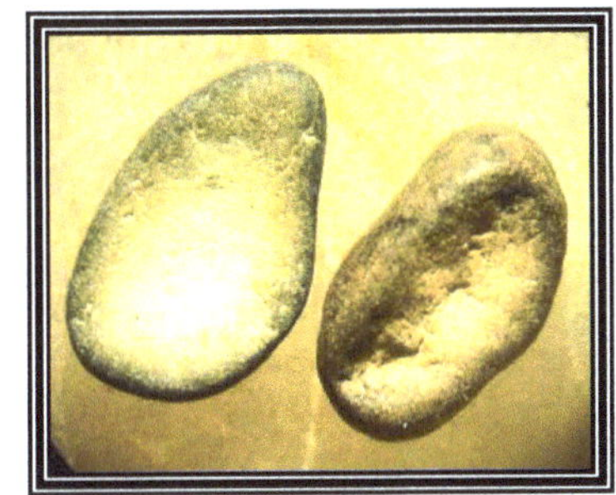

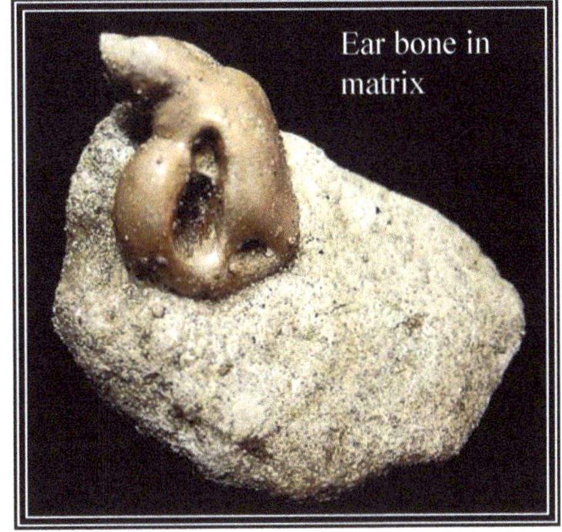

Sharks teeth in the Matrix

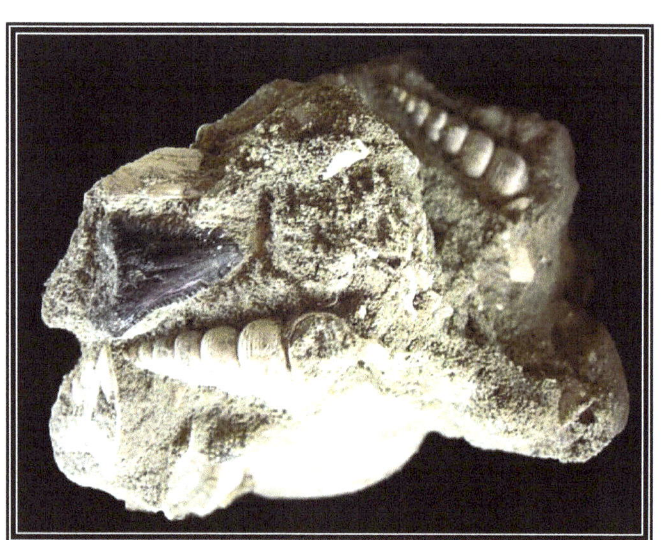

Nicely preserved (choice matrix specimens) Calvert Cliffs Maryland

Nicely preserved (choice matrix specimens) Calvert Cliffs Maryland

Calvert Beach and Matoaka, approximately 1 mile East of St. Leonard Calvert Co. Maryland

Alluvium covered cliffs, Matoaka Calvert Beach

Cliff face at Matoaka

Calvert Cliffs at Matoaka

Bone excavation CMM

View to the south

Calvert Cliffs at Matoaka

Calvert Cliffs at Matoaka

View to the north

Bone fragments in marl discovered by
friends in company along Matoaka beach

Working on bone material found at Matoaka

Un-identified bone appearing to be skull material

Calvert Cliffs at Matoaka

Scanning the stratigraphy

View to the south

Fossil coral bed in cliff face just south of Matoaka along Calvert Beach

Calvert Beach just south of stream running through property boundaries from Matoaka

View looking south

Fossil tree oysters in cliff with coral bedding (Melina)

3.5"   Brownies Beach March 2008
Kerry Matt
<stoneman07@netzero.com>

Brownies Beach Spring 2008 TL> fish vertebra, TR> mammal vertebra in matrix, MTL> porpoise tooth, LMTL> seven gill shark tooth, ML> Squalodon Tooth. MR> ray tooth, lower snaggletooth shark, Delphinodon tooth, LL> 3.5" Carcharodon Megalodon Tooth, LR fish tail vertebrae in matrix

www.ingramcontent.com/pod-product-compliance
Ingram Content Group UK Ltd.
Pitfield, Milton Keynes, MK11 3LW, UK
UKHW060119300726
14090UKWH00002B/260

* 9 7 8 1 4 5 0 0 1 5 5 1 6 *